ICME-13 Topical Surveys

Series editor

Gabriele Kaiser, Faculty of Education, University of Hamburg, Hamburg, Germany

More information about this series at http://www.springer.com/series/14352

Marcus Nührenbörger · Bettina Rösken-Winter
Chun-Ip Fung · Ralph Schwarzkopf
Erich Christian Wittmann
Kathrin Akinwunmi · Felix Lensing
Florian Schacht

Design Science and Its Importance in the German Mathematics Educational Discussion

Marcus Nührenbörger
Fakultät für Mathematik
Technische Universität Dortmund
Dortmund
Germany

Erich Christian Wittmann
Fakultät für Mathematik
Technische Universität Dortmund
Dortmund
Germany

Bettina Rösken-Winter
Didaktik der Mathematik
Humboldt-Universität zu Berlin
Berlin
Germany

Kathrin Akinwunmi
Fakultät für Mathematik
Technische Universität Dortmund
Dortmund
Germany

Chun-Ip Fung
Department of Mathematics and Information
    Technology
Education University of Hong Kong
Hong Kong
China

Felix Lensing
Freie Universität Berlin
Berlin
Germany

Florian Schacht
Fakultät für Mathematik
Universität Duisburg-Essen
Essen
Germany

Ralph Schwarzkopf
Fakultät V - Institut für Mathematik
Carl von Ossietzky Universität Oldenburg
Oldenburg
Germany

ISSN 2366-5947   ISSN 2366-5955   (electronic)
ICME-13 Topical Surveys
ISBN 978-3-319-43541-1   ISBN 978-3-319-43542-8   (eBook)
DOI 10.1007/978-3-319-43542-8

Library of Congress Control Number: 2016946300

Printed on acid-free paper

This Springer imprint is published by Springer Nature
The registered company is Springer International Publishing AG Switzerland

Main Topics You Can Find in This "ICME-13 Topical Survey"

- Roots and scope of design science;
- The role of substantial learning environments;
- Examples of current design research projects and developments;
- Collective teaching experiments;
- Commonalities and variations between design science and design research.

Contents

**Design Science and Its Importance in the German
Mathematics Educational Discussion** 1
1 Introduction 1
2 Survey on the State of the Art............................. 2
   2.1 Roots and Scope of Design Science 2
   2.2 Design Science Between Normative and Descriptive
        Approaches....................................... 10
   2.3 Developing Mathematics Teaching and Mathematics
        Teachers.. 18
   2.4 Collective Teaching Experiments: Organizing a Systemic
        Cooperation Between Reflective Researchers and Reflective
        Teachers in Mathematics Education 26
   2.5 Design Science and Design Research: Commonalities
        and Variations.................................... 34
3 Summary and Looking Ahead............................. 38
References 40

Design Science and Its Importance in the German Mathematics Educational Discussion

1 Introduction

Within the German-speaking tradition, considering "mathematics education as a design science" has been connected to the seminal work by Wittmann. In his famous lecture at ICME-9 in 2000 he underlined the role of substantial learning environments while elaborating on how mathematics education can be established as a research domain. From their very nature, substantial learning environments contain substantial mathematical content even beyond the school level and also offer rich mathematical activities for (pre-service) teachers on a higher level. Exploring the epistemological structure reflected in substantial learning environments or reflecting didactical principles while testing substantial learning environments in practice adds to a deeper understanding of both the mathematics involved and students' learning processes.

In view of his work on substantial learning environments, Wittmann not only elaborated on why mathematics should be the "core" of mathematics education, but also how the "related disciplines" can be centred on this core. In particular, he called for recognizing mathematics education as a scientific field in its own right, from where the constructive development of and research into the teaching of mathematics starts. Thus, mathematics education hllas been conceptualized as a constructive scientific discipline that has contributed teaching concepts, units, examples, and materials. The main objective has been to develop feasible designs for conceptual and practical innovations, involving the teachers actively in any design process.

The next chapter is dedicated to elaborating on these aspects more deeply and discussing how the theoretical orientations are reflected in Wittmann's design research project, mathe 2000. Having deeply explored the concept of design science, attention is then paid to present two examples of current projects that depart

© The Author(s) 2016
M. Nührenbörger et al., *Design Science and Its Importance in the German Mathematics Educational Discussion*, ICME-13 Topical Surveys,
DOI 10.1007/978-3-319-43542-8_1

from design research but also pursue particular accentuations: The first contribution (Sect. 2.2) reports on a project that positions design science between normative and descriptive theories. That is, mathematical learning opportunities have been designed from a normative viewpoint while learning processes have been explained on the basis of descriptive theories. Examples are provided that illustrate such a proceeding. The second contribution (Sect. 2.3) also reports on a developmental research project. Here, studying carefully designed teaching units is in the focus while paying explicit attention to problems that teachers face in the classroom. One main goal of the research is to help teachers to develop their mathematical knowledge and to ultimately enhance their teaching. Subsequently, in Sect. 2.4 Wittmann elaborates on the conception of empirical studies within design research that start from the mathematics involved in terms of structure-genetic didactical analyses. Particularly, bridging didactical theories and practice is pursued by collective teaching experiments. Finally, Sect. 2.5 takes up the strands presented so far and sheds light on developments of design science from a national and international perspective. Thereby, particular emphasis is assigned to design research from a learning perspective as extending characteristics from design research. The chapter ends with a summary and an outlook on further developments.

2 Survey on the State of the Art

2.1 Roots and Scope of Design Science

Kathrin Akinwunmi, Felix Lensing, Marcus Nührenbörger, Bettina Rösken-Winter and Florian Schacht

The fundamental scopes in mathematics education have been the initiation and the support of learning processes. Therefore, one important strand has been concerned with designing learning environments and exploring the induced learning processes. In the German-speaking tradition, the notion of *design science* is closely connected with Wittmann, who has emphasized and elaborated this concept over the past decades (cf. Wittmann 1995). The careful analysis of the mathematical substance and the potentials of mathematical structures within the specific designs have played a prominent role. In this regard, the concept of design science is embedded in the German *Stoffdidaktik* tradition. In a broader perspective, the approach has played a distinctive role within prominent European traditions concerned with designing and evaluating learning material such as Realistic Mathematics Education (RME, cf. Gravemeijer 1994) in the Netherlands or the Theory of Didactical Situations (TDS, cf. Brousseau 1997) in France. Against this background, the unique elements of mathematics education as a design science following Wittmann are presented in this section.

Nowadays, mathematics education seems to have quite naturally developed into a well-established discipline (Kilpatrick 2008). Still, in the 1970s the status of mathematics education as a distinct science and an academic field of its own right was questioned. Emphasizing that mathematics education needed a scientific and theoretically fundamental basis, Erich Christian Wittmann explicitly conceptualised mathematics education as a design science. Being trained in mathematics—Wittmann finished his Ph.D. in Algebra in 1967 and became a full professor in mathematics education in Dortmund, Germany, in 1967—Wittmann emphasized the fundamental role of the discipline in any conceptualisation of mathematics education as a field of study and practice. In his seminal paper about design science, published in *Educational Studies in Mathematics* in 1995, he pointed out the necessity "to preserve the specific status and the relative autonomy of mathematics education" (p. 355). Though mathematics education differed from pure mathematics, it was not simply a conglomerate discipline of mathematics with the related sciences of psychology, sociology, or pedagogy. Rather, it was characterized as an applied discipline whose core task was to develop practical, constructive products of acknowledged quality for teaching mathematics. Thus, this core encompassed many different components such as analysing mathematical activities and related mathematical thinking as well as developing local theories on mathematizing or problem solving, to name but a few (cf. Wittmann 1995). How the core of mathematics education was connected to related disciplines, areas, and fields is displayed in Fig. 1.

Wittmann (1995, p. 357) underlined in particular that "although the related areas are indispensable for the whole entity to function in an optimal way, the specificity of mathematics education rests on the core, and therefore must be the central component." In a recent interview[1] with junior scientists Kathrin Akinwunmi, Felix Lensing, and Florian Schacht, he deliberated on this point as follows:

> Recently, this graphic has been the centre of much discussion. It shows my humble attempt to give mathematics education a comprehensive framework in order to bring out its special character. However, it is particularly important that the graphic is not interpreted as something restrictive, as a boundary. No discipline should be pitched against another. Mathematics education is a tree that has roots and these roots must be cultivated. Nevertheless one must not lose one's way amongst the related disciplines, but must nurture the core. (Interview with Wittmann 2015, translated by the authors)

In 1974 Wittmann had already advocated that mathematics education should systematically examine "mathematics teaching" from the position of a mathematics teacher. The ultimate goal of empirical work has therefore not been neutral, descriptive, or normative, but rather constructive and prescriptive. The particular

---

[1]Erich Wittmann was interviewed by Akinwunmi, Lensing, and Schacht with on 4/12/2015 at TU Dortmund's IEEM as part of preparing the ICME 13 Thematic Afternoon on German-Speaking Traditions in Mathematics Education Research.

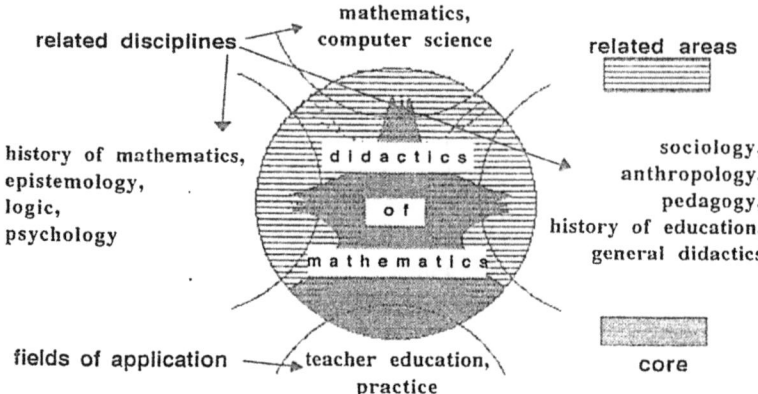

Fig. 1 The core of mathematics education in relation to other disciplines, areas, and fields of application (Wittmann 1995, p. 357)

task of mathematics education as a design science has been, on the one hand, to develop the best possible designs with respect to the respective curricular framework. On the other hand, the intention has also been to optimise how mathematics is taught in the classroom and to develop practical aids for teachers. During the interview, Wittmann thoroughly reflected on his view on mathematics education as a design science and pointed out the following:

> Perhaps to understand why it is so important to me, you have to remember where I come from. After completing my Ph.D. in Mathematics, everyone thought I would become a professor in mathematics, but then there was that fateful encounter with Freudenthal at a conference. I had previously studied didactics intensively, that was something that had always interested me, and after this encounter with Freudenthal it was clear to me that I should switch to didactics. But then I was faced with a particularly difficult situation, that of leaving an established science, the template for all sciences, and entering an undefined field. ... And that concerned me: What can be done to ensure that mathematics education becomes established. Then it occurred to me that there are indeed other disciplines, such as engineering, that are not exactly natural sciences. Yes, and I thought that mathematics education was something similar. And then I came across this book by Herbert Simon (1979), *The Sciences of the Artificial*. He was preoccupied with this very question. He confronted fields such as administration theory or computer science or economics and asked what do these represent in relation to the other sciences? ... And Herbert Simon had also found the solution. Mathematics education is a design science. Then it became clear to me: I do not know what other purpose mathematics education should have other than to improve teaching and the education of teachers. That is the ultimate goal. This must be the focus and the means of judging success. (Interview with Wittmann 2015, translated by the authors)

In the interview, Wittmann painted a broader picture of the discussion in the 1970s on what established mathematics education as a research field. Particularly, he elaborated on what has influenced him to conceptualise mathematics education as a design science. For Wittmann, improving teaching has been at the very heart of mathematics education. Fostering substantial mathematics teaching turns the attention to learning environments that satisfy specific demands with respect to the mathematics involved.

In 2005, Lesh and Sriraman still called for re-conceptualizing "the field of mathematics education research as that of a design science akin to engineering and other emerging interdisciplinary fields which involve the interaction of 'subjects,' conceptual systems, and technology influenced by social constraints and affordances" (p. 490). They found mathematics education to be "still in its 'infancy' as a field of scientific inquiry" (p. 490). In the interview, Wittmann elaborated on his viewpoint of introducing design science while referring to the work of Simon (1970), who distinguished between "artificial sciences" and "natural sciences" as follows:

> Historically and traditionally, it has been the task of the science disciplines to teach about natural things; how they are and how they work. It has been the task of engineering schools to teach about artificial things: how to make artefacts that have desired properties and how to design. (Simon 1970, p. 55)

In this sense, artificial sciences develop and study artificial objects that are in principle adaptable and can be changed with reference to specific objectives or functions. Mathematics education, according to Wittmann (1992), is thus a science that deals with the artificial objects of mathematics teaching (i.e., with exercise tasks as well as specific mathematics learning and teaching processes) and adapts these constructively with a view to further developing mathematics education.

2.1.1   The Role of Substantial Learning Environments

Within design science, developing appropriate *teaching examples* is the central task of mathematics education researchers. These teaching examples allow for integrating different aspects of the various related disciplines relevant to the teaching and learning of mathematics and for making them usable for designing mathematics teaching. From this perspective, structural issues of mathematics and an orientation towards applications are important. Students' learning is then considered to be an active process, assigning particular relevance to pedagogical theories and methods of social learning.

By teaching examples Wittmann (2001) does not mean complete and detailed lesson units. Instead, he is more interested in *substantial learning environments* that reveal to learners an individual space for error and discovery as well as their own paths of learning. The characteristics of such substantial learning environments are:

1. They represent fundamental objectives, contents, and principles of mathematical learning at a particular level.
2. They are based on fundamental mathematical content, processes, and procedures beyond this level and contain a wealth of mathematical problems ("exercises").

3. They can be flexibly tailored to the specific conditions of a particular class.
4. They integrate mathematical, psychological, and educational aspects of mathematics teaching and learning and therefore provide a rich field for empirical research (cf. Wittmann 2001a, p. 2).

Although not referring to the concept of substantial learning environments explicitly, the construction of "learning milieus" that are clearly conceptualized in delineation of sequences of exercises (Skovsmose 2011) has become a main focus of different strands in European mathematics education. The approaches have naturally differed according to their theoretical backgrounds and aims in each tradition. In France, both strands of the classical tradition of French "*didactique des mathematiques*" shared a clear emphasis on investigating "the genesis and the ensuing peculiarities of the (mathematical) 'contents' studied at school" (Chevellard and Sensevy 2014, p. 38). In the Theory of Didactic Transposition (TDT, cf. Chevallard 1991) as well as in the Theory of Didactical Situations (TDS, cf. Brousseau 1997), this has necessarily included the design of didactical material that is evaluated and further developed in an ongoing mediation process between educational practice and theory development. Although both theoretical strands further share the primacy of mathematics for the development of practical learning situations, learning as such has been conceptualized as a fundamentally social phenomenon. This has led to a specific methodological approach: learning situations have not been understood as epistemic relationships between a student and the object of learning but have rather been investigated as inter-subjective communication processes that are always embedded in institutionalized learning milieus (Brousseau 1997). These basic assumptions were systematically developed further in anthropological approaches towards the learning of mathematics such as the Anthropological Theory of Didactics (cf. Chevellard and Sensevy 2014), but it was the theory of Didactical Engineering (cf. Artigue 1994), also based on Theory of Didactical Situations and Theory of Didactic Transposition, that has come closest to the depicted understanding of mathematics education as a design science.

In the Netherlands, several studies (Gravemeijer 1994; Van den Heuvel-Panhuizen and Drijvers 2014) are also related to this paradigm (Margolinas and Drijvers 2015). The domain-specific instruction theory RME constitutes six core teaching principles that share some similarities to the concept of learning environments. These principles are (1) the activity principle, (2) the reality principle, (3) the level principle, (4) the intertwinement principle, (5) the interactivity principle, and (6) the guidance principle (Van den Heuvel-Panhuizen and Drijvers 2014, p. 522f.). It is important to note that even though teaching-learning trajectories of RME often refer to real-life, the "realistic" in RME simply demands that "the problems are experientially real in the student's minds" (p. 521). This can also include the experience of a formal mathematical activity (reality): "the transition from 'model of' to 'model for' involves the constitution of a new mathematical reality that can be denoted formal in relation to the original starting points of the students" (Gravemeijer 1999, p. 155).

Another similar though differently named idea can be found in Skovsmose's (2001, 2011) concept of "landscapes of investigation." While landscapes of investigation can consist of a learning milieu with (a) exclusive references to pure mathematics, Skovsmose (2011, p. 39ff.) proposes two further possibilities: landscapes of learning in a milieu with (b) references to semi-reality (e.g., a simulated context) and (c) real-life references (drawing on resources from students' lived experience). It is landscapes of learning of the latter kind that make up for the paradigmatic "learning environment" of Critical Mathematics Education.

Referring to this broad scope of different European traditions in mathematics education, it seems to be necessary to further clarify the role of substantial learning environments in the design process itself. In the interview, Wittmann explained explicitly why substantial learning environments are the relevant objects in design science:

> So what are the objects of this design science? Well, my suggestion is that these are learning environments. ... Because they have a practical relevance and, at the same time, provide a link to the other areas (related disciplines). (Interview with Wittmann 2015, translated by the authors)

In order to design substantial learning environments, Wittmann considers the mathematical substance of the exercise, or in other words the "epistemological structure of the subject", to be of particular importance. The latter provides the basis of composing mathematical activities for children at very different ages and levels. However, the mathematical substance does not appear as the systematic deductive structure of already "finished" mathematics. Rather, it recalls Freudenthal's ideas and provides an understanding of mathematics in relation to the learner's perspective on the subject. To explicitly distinguish this "genetic viewpoint on mathematics learning" from the pure mathematical viewpoint, Wittmann (1995) wrote *MATHEMATICS* in capital letters.

Besides subject matter aspects, Wittmann (2004) suggested 10 "didactical principles," providing a guide for assessing substantial learning environments (see Fig. 2). The 10 principles encompass organisational, epistemological, and psychological issues relevant for the teaching and learning of mathematics. Wittmann (2004) especially underlined the below-mentioned role of these didactical principles:

> The principles ... are important in evaluating and checking already existing learning environments at all stages of the design process. However, they are of no help for designing them. Design is a creative act whereby elementary mathematics and its applications provide the raw material. Therefore it is absolutely essential that professional designers cultivate their own mathematical activities. First-hand experiences with elementary mathematics are essential also for a second reason: Children's learning processes can only be understood and studied properly by researchers who themselves have thorough experiences with mathematical thinking processes. Design based on intimate knowledge of the subject matter is also important for distinguishing which elements of a substantial learning environment are determined by the subject matter and which ones offer options for revision. (Wittmann 2004, in Link 2012, pp. 59f.)

While the didactical principles, such as the epistemological structure of the subject, offer an essential orientation for evaluating learning environments, the

Fig. 2 Organisational, epistemological, and psychological principles to assess the quality of substantial learning environments (Wittmann 2004 in Link 2012, p. 59)

creative act of developing new learning environments ultimately remains not explicitly comprehensible and explainable. This depends mainly on the "constructive imagination" of the developer (Wittmann 1992). Nevertheless, in the interview, Wittmann points out with respect to the work of Wheeler (1967) and Fletcher (1964) that elementary mathematics is the very first source for designing learning environments.

> The second thing that one needs to strive for is an understanding of the children, for instance their level of knowledge and skills. And consequently, one also needs an idea of the curriculum. Sometimes too little attention is paid to the curriculum. Working towards a good curriculum is also important. (Interview with Wittmann 2015, translated by the authors)

According to Wittmann (2015), the following characteristics provide orientation for developing substantial learning environments: mathematical substance and richness of activities at different levels, assessment of cognitive demands, curricular fit (in terms of content and general learning objectives), curricular coherence and consistency, curricular coverage, exercise potential, and estimation of the time required. Mathematics education as a design science is therefore a creative science based on the solid knowledge of fundamental mathematical structures and processes combined with profound knowledge of children's learning, professional learning requirements, and objectives of mathematics teaching, also with reference to curricular frameworks.

During the interview, Wittmann makes clear that he appreciates the different attempts to promote mathematics education on an international level.

> Concerning the Japanese and Asian didactics for example, it was Jerry Becker who made me aware of the lesson studies that are also recognised in the USA by now. Mathematics education researchers in Japan have already worked according to design science principles for quite a long time but in their own interesting way. Another example is Guy Brousseau (Brousseau 1997); we have talked about design science very often. In his work some social components, which I pay less attention to, are very distinct.

Hence, there are differences in the role and the embedding of substantial learning environments in the international traditions in design science, mainly concerning the research focus and the way of developing and testing the environments. Considering the design principles above, in recent publications Wittmann (2015) speaks of a "structure-genetic didactical analyses," which he regards as "empirical research of the first kind." Such an empirical approach links material analysis with basic mathematical processes that occur during children's active occupation with mathematics, alone or in exchange with others, and demonstrate different approaches, forms of representation, errors, or misunderstandings. Significant publications in this field include the *Handbücher produktiver Rechenübungen* (2 volumes) and the textbook *Das Zahlenbuch* (Volume 1–4) (*The Book of Numbers*, used from kindergarten up to Grade 4).

2.1.2 Scope of Design Science with Respect to Curricular Embedding

While the methodological foundations of design science were put in place in the 1970s and early 1980s, Wittmann, together with his Dortmund colleague and friend Gerhard Müller, initiated the mathe 2000 developmental project in 1987, which has continued since 2012 under the name mathe 2000+. This project is science based and simultaneously relevant in practice. Wittmann and Müller developed and researched substantial learning environments that enable an active approach to mathematics within a meaningful context and support the students in developing understanding of mathematical structures.

The concept of mathematical learning processes and mathematical teaching processes has been strongly influenced by the work of Piaget, Kühnel, and Dewey. For example, Wittmann referred to Kühnel, who found in 1916 that "arithmetic teaching" should be directed to the activity of the student. In particular, Kühnel demanded that the student role should no longer be set to receive, but to develop actively.

With the mathe 2000 project, Wittmann and Müller aligned themselves with the curricular requirements of the 1985 curriculum (in North Rhine-Westphalia) and investigated its implementation empirically based on targeted teaching examples.

The research and development of mathematics teaching was based on the assumption that individual learning processes always depend on active discussion of mathematical issues and, at the same time, are also linked to social interactions where such discussions are realised.

The quality of the learning environments developed in the mathe 2000 project has been continuously reviewed by Wittmann and Müller in collaboration with teachers, so that the processing of learning environments underwent cyclical revisions. Empirical studies that methodically examine the implementation of learning environments and the associated teaching and learning processes represent, according to Wittmann (2015), empirical research of the second kind. Accordingly, Wittmann (1995) also speaks of empirical research centred around substantial learning environments. In this case, the learning environments serve as a basis to investigate more closely the mathematics teaching processes and, in particular, the mathematical thinking of students. Furthermore, such studies serve to revise and improve the learning environment. Echoing Piaget's clinical interviews, Wittmann (1995) proposed "clinical teaching experiments" as a suitable empirical method:

> As a result we arrive at "clinical teaching experiments" in which teaching units can be used not only as research tools, but also as objects of study. The data collected in these experiments have multiple uses: They tell us something about the teaching/learning processes, individual and social outcomes of learning, children's productive thinking, and children's difficulties. They also help us to evaluate the unit and to revise it in order to make teaching and learning more efficient. ... Clinical teaching experiments can be repeated and thereby varied. By comparing the data we can identify basic patterns of teaching and learning and derive well-founded specific knowledge on teaching certain units. (Wittmann 1995, p. 367f.)

2.2 Design Science Between Normative and Descriptive Approaches

Ralph Schwarzkopf

The main interest of design research in mathematics education is to influence the practice of teaching and learning mathematics from a theoretical point of view. Whereas some researchers try to modify curricular products, the present contribution concentrates on design research that focuses on the teaching and learning processes and on generating local theories (cf. Prediger et al. 2015, p. 881). Substantial learning environments play the main role, standing at the very centre of mathematics education as a design science (Wittmann 2001a, b). Hence, design research means to create learning environments from a normative perspective, trying to support the teachers to enfold mathematics in an authentic and substantial way. But how can we know whether the constructed learning environments really lead to fruitful learning opportunities in classroom activities? Of course, the researchers do not only design samples of tasks, they are also interested in their effects on the teaching and learning mathematics. For this it is very important to "investigate the process of learning, not only its inputs and outputs" (Prediger et al. 2015, p. 882); in other words, it is not satisfying to only measure *if* the intended learning goals are reached, the researchers should also build theories to explain *how* learning opportunities are realised within the mathematics lessons. These investigations are typically influenced by more descriptive theories of learning and teaching from scientific neighbour-disciplines of mathematics education such as pedagogy, sociology, or psychology.

Within the process of design research, the outcome of these empirical investigations influences the following planning of substantial learning environments. Hence, for these "iterative cycles of invention and revision" (Prediger et al. 2015, p. 879) the researchers have to switch between normative and descriptive points of view: In the phase of planning, they try to have an influence on the teaching and learning processes from a mathematical point of view and assume that the learning of the children is more or less a result of the realized substantial learning environment. In the phase of analysing data of the design experiments they assume that the social process of teaching and learning decides in a somehow autonomous way about the development of mathematics, i.e., the mathematical content of the learning process is a result of the social processes between teacher and students.

The following sections try to build bridges between these two approaches, i.e., to use this methodological contradiction in a productive way. I argue that concepts of argumentation processes are fruitful for both designing substantial learning environments from a normative point of view and explaining the observed learning opportunities within a descriptive interest of creating local theories.

2.2.1 Argumentation as a Design Principle

As we all know, the scientific discipline of mathematics is inseparably bound to mathematical proving: Whenever we make a new assumption, we search for verification by a mathematical proof. Hence, in mathematics education we stress the central role of proving in the teaching and learning mathematics, even within primary school (Wittmann 1998).

Of course, there is no exact definition for the concept *proof* within mathematics (cf. Heintz 2000; Wittmann and Müller 1988): Whether we accept something as a (good) proof or not depends on our requirements, i.e., on the main function of the proof within the special given situation. According to Hanna (2000, cf. Winter 1983), the main function of a proof, having in mind situations that provide the learning of mathematics, should be *explaining*: "In the educational domain, then, it is only natural to view proof first and foremost as explanation, and in consequence to value most highly those proofs which best help to explain" (Hanna 2000, p. 8).

The teachers, especially in primary schools, try to realise this function of proving by providing processes of argumentation, i.e., by introducing the pupils to developing arguments for mathematical assumptions. This approach leads to the special role of argumentation in teaching and learning mathematics, especially for design research. One of our main goals is to design examples for *learning opportunities* that provide mathematical argumentation from the beginning of school (or even earlier) in order to help the children understand the mathematical structures behind arithmetical phenomena. Operative proofs are a very fruitful source for designing learning environments that provide mathematical argumentation, especially (but not exclusively) in the primary level of school (Wittmann and Müller 1988; Wittmann 1998, 2001; Steinweg 2006, 2013).

In the following example we focus on a second graders' textbook (Figs. 3, 4 and 5), namely on the *Zahlenbuch* (Wittmann and Müller 2012). This particular page adapts operative proofs from the thematic field of figurate numbers, i.e., it provides the discovery of the main relation between square numbers and triangular numbers: The sum of two neighbouring triangular numbers results in the corresponding square number. For this, the pupils first learn about the figurate representation of square numbers (see Fig. 3) and have to calculate and sketch the next five square numbers, following the given representations of the first four.

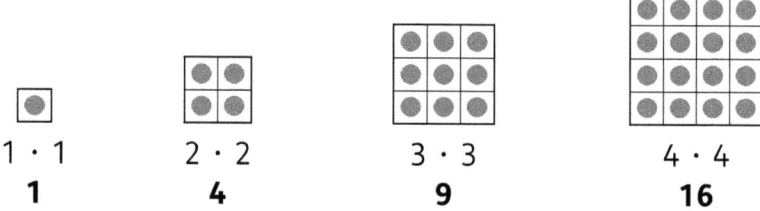

Fig. 3 Figurate representation of square numbers

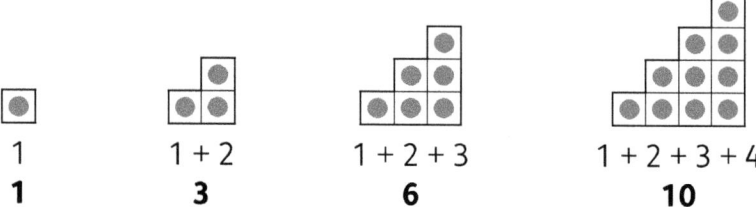

Fig. 4 Figurate representation of triangular numbers

Fig. 5 Sums of 2 following triangular numbers

Plusaufgaben von benachbarten Dreieckszahlen.

(a)	(b)
1 + 3	21 + 28
3 + 6	28 + 36
6 + 10	36 + 45
10 + 15	45 + 55
15 + 21	

In the second task (see Fig. 4), the children are confronted with arithmetic and geometric representations of triangular numbers, also having to sketch and calculate the next few examples. Afterwards, they are requested to add two neighbouring triangular numbers in an operative way (see Fig. 5). The children will surely discover the above-mentioned relation between the square and the triangle numbers.

Obviously, this relation stands for a very important and also very difficult learning goal in mathematics: It is very important to foster children's thinking about meaningful relations between the mathematical objects, i.e., the children have to construct an algebraic point of view about arithmetical structures. Hence, it is not the relation itself, but the understanding of its structure that is important for children's learning. In this case, as in many other situations, the pupils will have to understand why the two different concrete objects (the sum of two neighbouring triangular numbers and the corresponding square numbers) are equal from a mathematical point of view.

From a normative point of view one main goal for the children on their way from arithmetic to algebra is the ability to understand equalities as a theoretical relation between two empirically different objects (Nührenbörger and Schwarzkopf 2016): Within the given example, the calculations leading to the triangle numbers are completely different from the calculations that lead to the square numbers. Hence, these two kinds of numbers—from an empirical point of view—are not the same, but two different objects. Only from a more theoretical point of view one can see that one square number can be disassembled into two triangle numbers where one follows the other or, vice versa, that the sum of two triangle numbers where one follows the other equals one square number. For the children's learning of mathematics, this is a very difficult learning step: From a theoretical point of view, two objects can regarded as equal, although they are not the same from an empirical point of view.

2.2.2 Argumentation from a Descriptive Point of View

To understand this "miracle of learning" (Heuvel-Panhuizen 2003), "shift of view" (Hefendehl-Hebeker 1998), "modulation of framing" (Krummheuer 1995), or however one names these "fundamental learning" steps (Miller 1986), we need more descriptive theories, e.g., the epistemological theories of Steinbring (2005, pp. 214ff.). According to Steinbring, the learning of mathematics oscillates between two poles: On the one side there is the empirical situatedness that leads to the construction of factual knowledge on somehow visible objects and their handling. On the other side there is the relational universality, i.e., the construction of pure relations between the objects, represented only by abstract mathematical symbols such as algebraic formulas. A fruitful learning opportunity can only emerge when the construction of knowledge is balanced between these two poles of knowledge construction, in other words: Neither dealing with the concrete objects nor expressions of the pure relations can help the children on their way from doing the

Fig. 6 Relations between
square and triangle numbers

calculations to understanding their underlying algebraic structure, i.e., the equality
between the two mathematical objects.

The textbook offers the possibility of constructing a balanced knowledge in this
sense: According to Winter (1982), understanding an equality between two mathematical terms means understanding that the terms are different representations of
the same mathematical object. In Fig. 6 one can see two children showing the key
to this flexible interpretation: One can see both the two neighbouring sums of the
first natural numbers *and* the square number of the biggest summand in the one
object that the children hold in their hands.

How can we help the children on their way through the fundamental learning
steps? According to Miller (1986, 2002), children in primary schools are rather
social than autonomous learners, i.e., they are not able to construct this kind of new
knowledge completely individually.

Hence, at least for children in the primary level, interaction processes are necessary to realise fundamental learning processes. However, as we all know, interaction processes are strongly influenced by social conditions. In short, social
routines can dominate the negotiation of mathematical meanings (Voigt 1994). In
these cases the interaction processes are only little helpful for the learning of
substantial mathematics.

To realise fundamental learning steps we need special kinds of social processes,
namely situations of collective argumentations: The children must be confronted
with a problem that makes it somehow impossible for them to go on by routine and
they first have to solve that problem in an argumentative way (Miller 1986;
Schwarzkopf 2003). Unfortunately, according to Miller, argumentation is a very
stressful kind of interaction so that people often try hard to solve their problems in a
non-argumentative way. Even within mathematics classrooms, where argumentation is one of the learning goals, there are many opportunities to avoid argumentation—at least one can always ask the teacher as an expert in mathematics
(Schwarzkopf 2000). Hence, the mathematical content alone is not enough to initiate collective argumentation. In addition, the social condition of the classroom is a
very important factor.

2.2.3 Example: The Realisation of Argumentation

Within the project PEnDEL M (practice-oriented development projects in discussion with educators and teachers by Nührenbörger and Schwarzkopf), we use both the normative and the descriptive meanings of argumentation. For this, we started a variety of different teaching-learning experiments such as group work, peer interviews, and whole-class instruction with the goal to initiate substantial argumentation in mathematics lessons. In the revision of our experiments, we then try to understand the social conditions that support the emergence of collective argumentation in order to learn how to initiate argumentation in subsequent experiments in a more effective way. Thereby we are influenced by learning and teaching theories from scientific neighbour-disciplines of mathematics education, mainly using approaches of symbolic interactionism and ethnomethodology (Voigt 1994; Krummheuer 1995; Yackel and Cobb 1996), theories of argumentation (Schwarzkopf 2003), and epistemological theories (Steinbring 2005; Nührenbörger and Steinbring 2009).

In this example, a fourth grader's class discussed triangle numbers. Because the pupils had no experiences with figurate numbers, they understood the triangle numbers at the beginning of the experiment more or less as a new calculation procedure. Hence, the main goal of the experiment was to discuss the triangle numbers from a more algebraic point of view: The pupils were to get insight into the well-known relation $(1 + 2 + \cdots + n) = n \cdot (n + 1)/2$ for every natural number n.

The teacher, of course, did not develop this formula in the symbolic representation. She did it within the figurate representations and on the base of examples, comparing the long sum with the quite short multiplication of two neighbouring natural numbers (for a discussion of the whole experiment cf. Nührenbörger and Schwarzkopf 2010).

However, the most interesting episode occurred at the very beginning of the experiment. The children have already learned how to represent the sum of the first numbers as a triangular figurate set of dots. They also have discussed how to calculate a triangle number from the one before it. Both representations were given on the blackboard (see Fig. 7) and the whole class calculated the first 10 triangle numbers.

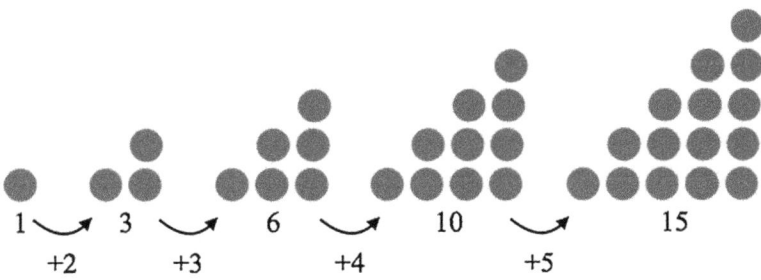

Fig. 7 Relations between triangular numbers

To motivate the children to have a more algebraic view on the triangle numbers, the teacher confronted them with a very long calculation so they had to calculate the 20th triangle number, and most of the children felt quite uncomfortable with this task. As the teacher asked the children for their calculations, David offered his description in the following way:

David: I just kept on calculating. I did plus 6 plus 7 plus 8 plus 9 plus 10 and then I was at 55 and I kept on calculating and then, after I don't know how much time, the result was 210

As we can see, David acted as the teacher expected: He calculated from the beginning on, following the given procedure and realising the awkwardness of the calculation; in other words, he produced the factual knowledge about the 20th triangle number completely within the empirical situatedness.

Other children of the class already tried to make the calculation shorter than David did:

Peter: Well, I have taken the pattern of the 10 and then I thought plus 9 plus 8 plus 7 plus 6 plus 5 plus 4 plus bla bla bla and so on and exactly like this I did it with the 20th pattern. I didn't paint it, I just thought it. And then I had the double of it. Because 20 is the double of 10 and so the result must also be the double

Peter's offer shows a more theoretical point of view on the calculation. Briefly speaking, he states proportional structures between the objects to avoid long-term calculations: According to him, the double index of the triangle number leads to the double number.

Hence, there were (at least) two different solutions of the problem: Some of the children had 210 as the result, while others followed Peter's idea and had 110 as the result. Although the second group thought that Peter's idea was somehow "logical," they could not explain it. According to Steinbring, Peter's assumption belongs to the children's relational universality: For them, it is very difficult to bind the rule to the given objects, as it is not related to the given problems. Maybe this is the aspect that made the children somehow excited; the interaction was more like a fight than like an argumentation. In this situation, Michael came up with the following argument:

Michael: Hello? You forgot the last tens. Just calculate 10 plus 11 plus 12 plus 13 plus 14 plus 15 plus 16 plus 17 plus 18 plus 19 plus 20

At this moment neither Michael nor the teacher could calm the class and the lesson was ended. The pupils were given the recalculation of the 20th triangle number as homework so that they could determine the right solution and restart the discussion the next day in a more rational way.

$$1+2+3+...+10+11+12+13+...+20$$
$$=(1+2+3+...+10)+(11+12+13+...+20)$$
$$=(1+2+3+...+10)+(10+1+10+2+10+3+...+10+10)$$
$$=2\cdot(1+2+3+...+10)+10\cdot10$$

Fig. 8 Twentieth triangle number

In fact, the next day Michael could explain his idea in a more fruitful atmosphere:

Michael: I first doubled the 55, because the 11 has got a 1, the 12 has got a 2, the 13 has got a 3 and then it has to be the same except the tens. And because there are 10 of them one can take 10 tens. So, 10 times 10 is 100 and then you know that it must be 210

Michael's idea seems to fulfil the conditions for fundamental learning steps: It is on the one hand bound to the triangle numbers (to their symbolic representation rather than to the figurate ones), but on the other hand it changes the view on the calculation steps to a more algebraic understanding (see Fig. 8).

So at least the highly emotional discussion on several solutions led to a very fruitful mathematical learning opportunity. At first the children had different calculation results and nobody could find a mistake in one of the solution methods. Then, due to this *social irritation*, the interaction could not go on in a routinized way, so there was a need for a collective argumentation—starting on an emotional rather than a rational level, but ending in a very fruitful learning opportunity.

The given example became paradigmatic in our project: In every substantial learning environment we search for opportunities to start *productive irritations*, i.e., for conditions that disturb the social routine and thereby may initiate collective argumentations. This approach is based on Piaget's (1985) work on *cognitive* conflicts (cf. Nührenbörger and Schwarzkopf 2016). According to Piaget, children can come to new ideas when they are confronted with contradictions of their individual expectations. If the existing cognitive schemas resist a child's possibilities of assimilation, there is a need to generate a cognitive consensus, i.e., a learning process starts. According to Miller (1986), the cognition can only be influenced in this way by disturbing social routines. Hence, in total, we work on tasks that help to generate fruitful social conditions for at least cognitive learning processes.

2.2.4 Closing Remarks

This chapter discussed the relations between the normative and descriptive perspectives within design research. For this, argumentation can be seen as a possible bridge between the constructive and reconstructive backgrounds of design science. For the research and design within mathe 2000, argumentation has been taken as somehow synonymous for proving, i.e., as one of the most important activities in mathematics that has to play a prominent role in the teaching and learning of mathematics from the beginning on. The intention of design projects such as PEnDEL M has not been primarily to invent more and more substantial learning environments. It has been to use the knowledge about social (micro)-processes in order to enrich the existing learning environments by fruitful social conditions for fundamental learning steps. Hence, our concept of argumentation has also been flanked by the ideas of several neighbour disciplines of mathematics education, especially the prominent foundations of interpretative research. In this way, we hope to use the well-known contradiction between normative and descriptive approaches to mathematics education in a productive way.

2.3   Developing Mathematics Teaching and Mathematics Teachers

Chun-Ip Fung

Traditional views about the division of labour between educational researchers and teachers have rested on the assumption that researchers are in a better position to develop and test theories, while teachers are responsible to apply, during their practice of teaching, whatever researchers have developed. The inadequacy of this model has brought about the research-practice dilemma (Kennedy 1997; Heid et al. 2006). In the mathematics education research community, "only 8 % said they do not believe their work had any impact at all. … Those who declare that their work did have an impact (45 %) use qualifiers such as *some, certain, little, limited*" (Sfard 2005, pp. 406–407, [italics original]).

Design-based research is now commonly regarded as a bridge between theory and practice (Design-Based Research Collective 2003; Ruthven et al. 2009). Carefully designed teaching units should therefore become the core objects of study of research in mathematics education. On the high end, these carefully designed teaching units have been called, according to Wittmann (1995, 2001a), substantial learning environments, which cover curriculum objectives at specified level, open up various possibilities for students to acquire important mathematical contents or abilities that go beyond that level, are flexible for adaptation to different classroom conditions, and integrate different aspects of mathematics teaching in such a way as to offer ample opportunities for empirical research.

Carefully designed teaching units alone do not constitute good teaching. That professional knowledge and competence of teachers are other important constituents needs no explanation. Research results have indicated that mathematics knowledge needed for good mathematics teaching is different from the academic mathematics commonly possessed by mathematicians (Adler and Davis 2006; Ball and Bass 2000; Davis and Simmt 2006). It follows that setting coursework requirements for mathematics teacher preparation or certification (see, for example, Conference Board of the Mathematical Sciences 2001; Leitzel 1991) may not ensure teachers are well prepared for the implementation of good instructional designs. In addition to creating carefully designed teaching units, there should also be measures to uplift the professional capabilities of teachers in order that they feel confident—and are competent—to implement these teaching units.

> The design of substantial learning environments around long-term curricular strands should be placed at the very centre of mathematics education. Research, development and teacher education should be consciously related to them in a systematic way (Wittmann 2001, p. 4).

In this chapter, I will illustrate briefly how I carry out design-based research and teacher development in an intertwining way through the curriculum topic of factors and multiples. Driven and limited by various local factors such as school curriculum, parents' expectation, public education policy, etc., how my work has been conceptualized, though following closely the German tradition exemplified by Wittmann's approach of "design science," may have caused it to have a different emphasis that rests upon values and priorities of Chinese communities.

2.3.1 Developmental Research in Action

Teaching for Mathematising (abbreviated TFM hereafter) is a developmental research project that I have been conducting in Hong Kong since 1998. It rests upon the following operational principles: (1) Research questions should come from problems relevant to the daily work of teachers. (2) Research work should aim at generating solutions for these problems in a local setting, mostly in the form of carefully designed teaching units or, at the high end, substantial learning environments. (3) Research output should attempt to address both the practical problems faced by teachers and the mathematical knowledge necessary for teachers to become confident and competent to implement the solutions. In short, research work of TFM has two foci. The first is to study and produce carefully designed teaching units for the purpose of solving problems faced by teachers. The second is to develop teachers' knowledge of mathematics and its teaching, as well as teachers' capability to act professionally.

How the work of TFM is carried out resembles, to some degree, the French approach of didactical engineering. According to Artigue (1994), constraints of teaching are epistemological, cognitive, or didactical by nature. Before a teaching unit, or even a part of it, is designed, the most important and significant work for the didactical engineer is to carry out an epistemological analysis of content structure. This analysis aims at identifying both the mathematical structure of a teaching unit, called its internal structure, and how the contents of the unit relate to other learning experience within the local curriculum and beyond, called its external structure.

This part of the work, if done well, serves to establish continuity between mathematics as a school subject and mathematics as an academic discipline, thus avoiding their possible divorce as highlighted by Watson (2008). Once a thorough epistemological analysis of content structure is done, the didactical engineer is ready to proceed to the engineering phase during which an outline of a teaching unit, or sometimes called a backbone design, is produced. A backbone design includes only the framework for the organization of the teaching unit and its major features, leaving the fine details for the teacher actually implementing it to fill in.

The main challenge for the didactical engineer is not just to produce backbone designs, but rather to produce backbone designs that are on one hand both mathematically and didactically sound, and on the other hand within reach of teachers with an average mathematical exposure. It would be of little practical value to produce designs the implementation of which requires a mathematical background comparable to that of George Pólya or Hans Freudenthal. Equally, designs that every teacher is comfortable to implement are likely to be those commonly found in classrooms where learning or teaching problems originate.

> If we design teaching Materials and methods, we should not only weigh up what can be learned and is worth learning, we should also be concerned about what kind of subject matter the teacher can learn to teach, or rather what we can teach our teachers to teach their pupils. (Freudenthal 1973, p. vii)

In essence, the didactical engineer is looking for solutions to learning or teaching problems, very often in the form of a backbone design, that an average teacher can learn how to implement.

In what follows, I will illustrate, based on the work of TFM on the teaching of factors and multiples, how practical problems of teaching are turned into the research work of a developmental researcher who takes up the role of both a didactical engineer, whose major concern is to generate backbone designs that could facilitate the solution of practical problems, and a teacher educator, whose major concern is to develop the professional knowledge and competence of teachers.

Hong Kong pupils learn factors and multiples in Grade 4 (at about 9–10 years of age). The official curriculum requires that pupils be able to find all factors of a positive integer by exhaustive listing. Figure 9 shows the U method, commonly found in local classrooms, for finding all factors of a positive integer. After exhaustive listing of all expressions of 12 as a product of two positive integers, pupils will follow the big U sign to list all the factors of 12, viz. 1, 2, 3, 4, 6, and 12. Many teachers and parents have reported that children have often had problems with this method. Either they did not understand what was actually going on or they failed to secure some of the expressions.

When translated into a mission for the didactical engineer, the problem is two-fold: one involves the meaning of the "U" and the other is how to get all the expressions. Given that both the accompanying arithmetic expressions and dot arrays suggest that the task has something to do with multiplication, it is not surprising to find pupils being hindered by attempting to find factors through

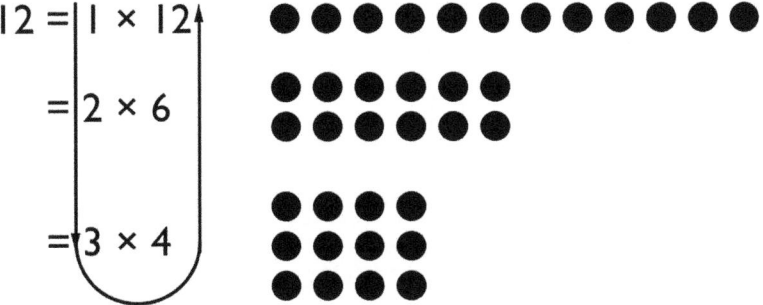

Fig. 9 The U method commonly found in Hong Kong classrooms

multiplication. With larger numbers such as 45, apart from the trivial expression $45 = 1 \times 45$, only $45 = 5 \times 9$ can be found in the familiar single-digit multiplication table. As a result, pupils will miss the expression $45 = 3 \times 15$ because this is not covered in the multiplication table they are asked to memorize. To rectify the situation, the teaching design should tie the concept of factor to the process of division, although factor relationship is commonly expressed in the form of multiplication. In addition, there should also be a part of the teaching that involves a way for the U method to evolve in a natural and convincing way.

Inspired by Euclid's approach to factors and multiples (see Heath 1956), the didactical engineer comes up with the following solution: Use identical squares in a row, called the number bar, to represent positive integers. Positive integer B is said to be a factor of positive integer A if its corresponding number bar measures that of A. In other words, when doing the measurement activity, using B to measure A can have two different possibilities. If Fig. 10 is the result, we say that B is a factor of A. Otherwise, Fig. 11 is the result and B is not a factor of A.

With a little discussion, pupils could point out that the phenomenon in Fig. 10 corresponds to the situation where positive integer A is divisible by positive integer B while the C in Fig. 11 corresponds to the positive remainder when A is divided by B. In this way, checking whether B is a factor of A is translated into the

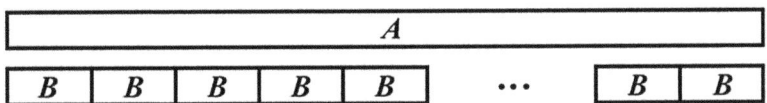

Fig. 10 The case when number B measures number A

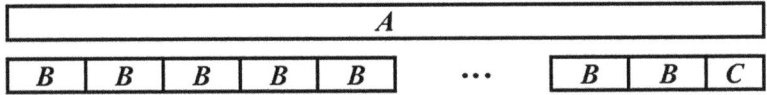

Fig. 11 The case when number B does not measure number A

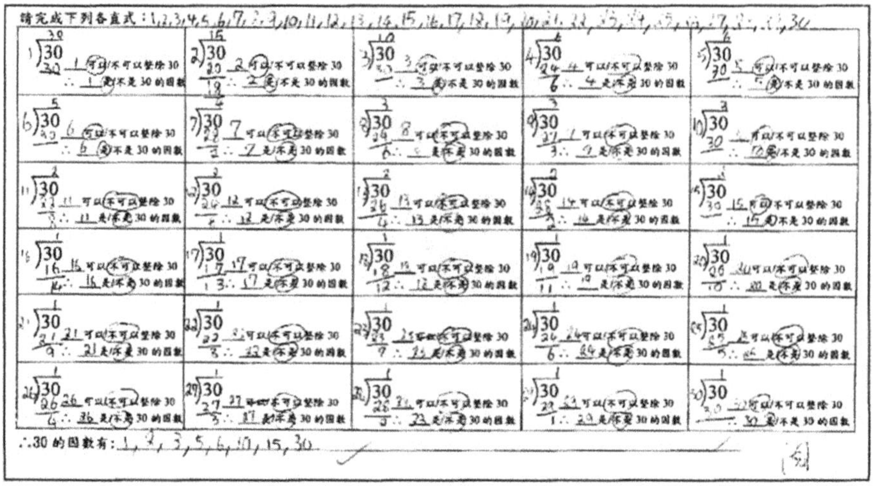

Fig. 12 Exhausting all factors of 30 by division

process of dividing *A* by *B*. Consequently, the concept of factor is tied to the division process.

Once division is confirmed to be a proper process to identify factors, pupils are asked to determine all the factors of a positive integer, say 30, by exhaustive division. Figure 12 (excerpt from Chow 2006) is an example of pupils' work in which the pupil did all the 30 calculations, with divisor running from 1 through 30, and circled to indicate if the divisor indeed divides 30 and hence is a factor of 30. By going through the tedious computation of dividing 30 by 1 through 30, pupils compiled a full list of factors of 30.

During the process, almost every pupil would have loved to cut down the number of calculations. The teacher demanded justification if pupils wanted to skip any of the division calculations. The same tedious work happened again when pupils were asked to find all the factors of 48. Feeling the pressing need to avoid a total of 48 division calculations, some pupils began to invoke whatever was applicable in their mind. Some pupils applied divisibility tests for 2, 5, and 10 (compulsory curriculum content) or even 3, 4, 8, and 9 (enrichment curriculum content). Subsequent discussion revolved around how to get the full list of factors without actually doing the exhaustive calculations.

The first milestone was the discovery that any proper factor cannot exceed half of the dividend. This was observed and explained by most of the pupils. Afterwards, with proper guidance of the teacher, pupils were able to realize that every time the remainder was zero and the quotient was not the same as the divisor, one actually found a pair of distinct factors.

Thus when doing the series of calculations in increasing order of the divisors, once a divisor was found to have appeared as a quotient in a previous calculation with zero remainder, the tedious work could stop because the full list of factors must have already surfaced. By listing two-factor factorization of a positive integer, with the first factor in increasing order, the U method of finding all the factors of the whole number evolved in a natural and convincing way (Chow 2006).

The adoption of this number bar measurement approach by the didactical engineer does not follow solely from just the need to find all factors of a positive integer. By the same visual demonstration, the study of common factors, multiples, and common multiples can be accomplished without much additional work. Figure 10 represents A being a multiple of B while Fig. 11 shows an example where this is not the case. That any common multiple of A and B is also a multiple of the lowest common multiple of A and B becomes obvious upon doing the number bar measurement. For more capable pupils, this number bar measurement sets the foundation for the introduction of a Euclidean algorithm for finding the highest common factor of two positive integers.

Indeed, careful examination of Fig. 11 reveals that whether a factor of B is also a factor of A depends precisely on whether it is a factor of C. It follows that common factors of A and B are exactly those common factors of B and C, enabling reduction of the problem of finding common factors of positive integers A and B to finding common factors of a smaller pair of positive integers B and C. Repeating the technique in a finite number of steps, one should ultimately find that one particular positive integer in some reduced pair is a factor of the other. Apparently, the common factors of the original pair of positive integers are exactly the factors of this particular positive integer, which in itself is the highest common factor. Not only could the algorithm thus found enable calculation of the highest common factor, it also explains that any common factor of A and B is also a factor of the highest common factor of A and B. Although this understanding goes far beyond curriculum requirement, it did happen in one classroom where pupils successfully learned the Euclidean algorithm. To go even further, that the product of two positive integers is equal to the product of their lowest common multiple and highest common factor can also be deduced without the notion of prime number (Fung 2004).

The Euclidean process mentioned above will not become a never-ending infinite process because a strictly decreasing sequence of positive integers has 1 (a factor of every positive integer) as a definite attainable lower bound. If any of A and B is relaxed to a positive number that is not an integer, the process ends in a finite number of steps when and only when A and B are commensurable. For the teacher, it is important to note that the side and the diagonal of a square have incommensurable lengths, and so have the circumference and the diameter of a circle.

2.3.2 Re-conceptualizing the Scientific Method

Teaching is carried out in a complex environment in which teachers are both agents of intervention as well as subjects who are likely changed by their environment. Once a new method brings about new teaching insights among teachers, the knowledge and beliefs of the teaching profession will change, which in turn causes a change in judgment and evaluation of teaching. For instance, the breakthrough caused by the number bar measurement mentioned above changed the beliefs of the profession to the extent that what used to be considered impossible is now commonly agreed to be attainable upon careful implementation. Research on teaching is thus being carried out by people who belong in the environment being studied, be it the classroom or the teaching profession. Attempting to change the teaching profession by contributing teachers within the profession resembles to some degree the concept of collective experiment described by Latour (2011). Teaching experiments, being socio-technical by nature, encompass far too many uncontrollable factors and parameters that cannot be properly accounted for. St. Clair (2005) called these superunknowns:

> Superunknowns are not just unaccounted for by present methods; they are simply unaccountable by their nature. The implication for educational research is a need to move away from the ideal of generalizable propositions and toward greater recognition of the value or educators' judgment (St. Clair 2005, p. 437).

Instead of being tied to generalizable propositions, TFM studies teaching through accumulation of reference cases, each of which is documented by one or a few of the teachers involved or sometimes in collaboration with the developmental researcher. Each case report serves as a reference point against which teachers' work can be studied and evaluated. It describes and analyses the work of one or a few specific teachers, in one or a few specific classrooms, with one or a few specific groups of pupils. These reports do not tell what can be observed in average lessons conducted by average teachers. Instead, they set forth attainable goals and describe possible scenarios that can be expected when certain backbone designs are implemented carefully by well-prepared teachers. This predominantly qualitative approach to the study of mathematics teaching looks for meanings that would enlighten practitioners instead of statistical results that aim at convincing people not belonging to the community of practitioners.

> Qualitative research is gaining ground, but we are still far from the point where mathematical methods can add a finishing touch to qualitative knowledge, and many researchers are even farther from the insight that mathematics is not able to do more than just this (Freudenthal 1991, p. 152).

The instructional approach to the teaching of factors and multiples outlined in this chapter was developed based on careful epistemological analysis of content structure and familiarity with contextual parameters. The former may be taken as research into mathematics itself, not in the form of pushing forward the frontier of the discipline, but in the form of defining, articulating, or sequencing the didactical contents in such a way as to better facilitate the teaching and learning of the discipline in the long run. It begins with activities commonly found in classrooms, then relates to some mathematical contents or values that go beyond school mathematics, and finally crystallizes into ideas on how teaching can be organized. Wittmann (2015) called it structure-genetic didactical analysis. The latter, unlike the former, is relatively more localized and hence may not bring about insight into the practice of teaching in the rest of the world.

In addition to setting attainable goals, the reference cases also point to possible directions for teachers to learn to strengthen their capability to teach professionally. In order to empower teachers to execute the designs in ways that are both mathematically and didactically sound, publications on TFM necessarily include in-depth elaboration of content structure based on advanced contents and/or processes of mathematics.

Fung (2008) explained division with remainder in a unified way. Irrespective of whether a and b are positive integers, fractions, or decimal numbers, once we require bq to be the closest integral multiple of b not exceeding a, we can get the unique expression $a = bq + r$, where q is a non-negative integer called the quotient, and $r = a - bq$ is the remainder, which may or may not be an integer and which satisfies $0 \leq r < b$.

In sum, TFM seeks to improve the teaching of mathematics by improving the processes by which mathematical knowledge evolves in the classroom. This can only be achieved by developing good instructional designs and preparing teachers for their implementation. By keeping abreast of classroom reality, teacher competence, and other related contextual parameters, the major work of the didactical engineer is to study and create designs. Once a design is in place, the major work of the teacher educator is to equip teachers with the knowledge and skills necessary for its implementation. As a developmental researcher, I have played the dual role of both a didactical engineer and a teacher educator in the past eighteen years. The collective effort of the whole TFM community has been put into continuously documenting both the design work and the implementation work. Because publications were written for, and often by, teachers, they were all in Chinese. They evidence the participation and contribution of teachers as practitioners. Their major purpose has been to assist teachers to learn some mathematics in order to be able to teach better. Video recordings of classroom implementation have been collected as far as possible to facilitate teachers to learn from each other.

2.3.3 Closing Remarks

The core value of developmental research lies in its contribution to initiate change for the betterment of teaching. To this end, contribution has to be made simultaneously to creating designs, improving designs, and uplifting teachers' mathematical knowledge. None of these can be achieved without going deeply into mathematics itself. Wittmann (2015) called this approach "mathematics education emerging from the subject." Eighteen years of TFM have confirmed unambiguously that by making carefully designed teaching units or substantial learning environments the core object of study for teachers as well as researchers, positive changes in classroom teaching can then be expected and prepared for. In addition, in order for teachers to play active roles as both a learner and a contributor in the endeavour, research publications in design science should meet as many as possible the following criteria: (1) They have direct relevance to the work of didactical engineers, teacher educators, or teachers. (2) They add new and improved understanding of mathematics that is conducive to creating better didactical processes in classrooms. (3) They add to teachers' repertoire of good teaching units. (4) They attract, engage, and change teacher readers.

2.4   Collective Teaching Experiments: Organizing a Systemic Cooperation Between Reflective Researchers and Reflective Teachers in Mathematics Education

Erich Ch. Wittmann

The success of any substantial innovation in mathematics teaching depends crucially on the ability and readiness of teachers to make sense of this innovation and to transform it effectively and creatively to their context. This refers not only to the design and the implementation of learning environments but also to their empirical foundation. Empirical studies conducted in the usual style are not the only option for supporting the design empirically. Another option consists of uncovering the empirical information that is inherent in mathematics by means of structure-genetic didactical analyses. In this chapter, a third option is proposed as particularly suited to bridge the gap between didactical theories and practice: collective teaching experiments.

In the following, five main points—as the headers of the sections—will describe in a nutshell the line of argumentation that is developed in this chapter:

- Mathematics education as a "systemic-evolutionary" design science
- Taking systemic complexity systematically into account: lessons from other disciplines
- Empowering teachers to deal with systemic complexity as reflective practitioners
- Collective teaching experiments: a joint venture of reflective researchers and reflective practitioners
- The role of mathematics in mathematics education.

2.4.1   Mathematics Education as a "Systemic-Evolutionary" Design Science

The proposal to consider mathematics education as a design science in Wittmann (1995) was stimulated by the intention to establish a sound methodological basis for a science of mathematics education that would guarantee a firm link between theory and practice and preserve the mathematically founded work achieved in curriculum development and teacher education by mathematics educators in the past (see Sect. 2). This proposal was based on the seminal paper by Simon (1970) in which the design sciences were characterized as being concerned with the construction of artefacts that serve defined purposes. In the design science mathematics education these artefacts are substantial learning environments.

Therefore, the core of this discipline consists of the design, the empirical investigation, and the implementation of substantial learning environments both with respect to boundary conditions set by society and beyond these constraints. It is obvious that there is a basic difference between design sciences, such as mechanical engineering and computer science in which artefacts (cars, computers, etc.) are developed that function according to natural laws in a completely controlled way and can be easily applied by the users, and design sciences such as economics, medicine in which the artefacts (marketing strategies, therapies, etc.) cannot take account of all elements of the environment in which the artefacts are to be used as this environment is simply too complex and also fluid.

Following Malik (1986), these two classes of design science can be distinguished as "mechanistic-technomorph" and "systemic-evolutionary" design sciences. Obviously, mathematics education belongs to the latter class for which a sharp separation between researchers and developers who design artefacts and users who simply apply them is not appropriate. The consequences for mathematics education have been indicated already in Wittmann (1995) and further elaborated in more general terms in Wittmann (2001). In the following, the practical implications of this systemic principle are discussed.

2.4.2 Taking Systemic Complexity Systematically into Account: Lessons from Other Disciplines

In the comprehensive literature in which appropriate models for the cooperation between researchers and practitioners in systemic-evolutionary design sciences are developed, Donald Schön's research on the "reflective practitioner" stands out in depth and in scope (Schön 1983). Schön was mainly concerned with management, architecture, psychotherapy, town planning, and those parts of engineering in which social aspects matter. Later he extended his analyses also to education (Schön 1991). This stimulated other educators to expand on them (cf., for example Wieringa 2011).

Schön describes the traditional relationship between "professionals" and "clients" as follows (Schön 1983, p. 292):

> In the traditional professional-client contract, the professional acts as though he agreed to deliver his services to the client to the limits if his special competence. ... The client acts as though he agreed, in turn, to accept the professional's authority in his special field [and] to submit to the professional's ministrations.

> In some parts of some practices ... practitioners can and do make use of the knowledge generated by university-based researchers. But even in these professions. ... large zones of practice present problematic situations which do not lend themselves to applied science. What is more, there is a disturbing tendency for research and practice to follow divergent paths. Practitioners and researchers tend increasingly to live in different worlds, pursue different enterprises, and have little to say to one another.

Schön replaces the unproductive traditional roles of researchers and practitioners with a picture in which the responsibilities are to some extent shared. Researchers act as "reflective researchers" and practitioners as "reflective practitioners" (Schön 1983, p. 323):

> In the kinds of reflective research I have outlined, researchers and practitioners enter into modes of collaboration very different from the forms of exchange envisaged under the model of applied science. The practitioner does not function here as a mere user of the researcher's product. He reveals to the reflective researchers the ways of thinking that he brings to his practice, and draws on reflective research as an aid to his own reflection-in-action. Moreover, the reflective researcher cannot maintain distance from, much less superiority to, the experiences of practice. … Reflective research requires a partnership of practitioner-researchers and researcher-practitioners.

However, Schön is far from denying researchers a special status: "Nevertheless, there are kinds of research which can be undertaken outside the immediate context of practice in order to enhance the practitioner's capacity for reflection-in-action" (Schön 1983, p. 309).

Schön distinguishes four types of this "reflective research" (Schön 1983, p. 309ff.):

Frame analysis: This type of research deals with general attitudes that provide practitioners with general orientations for their work.

Repertoire-building research: The focus here is on practical solutions of exemplary problems ("cases") that provide guidance not only in routine cases but also when it comes to dealing with similar new problems.

Research on fundamental methods of inquiry and overarching theories: This type is closely connected to both types mentioned above. It is directed to developing "springboards for making sense of new situations" for which no standard solution is available.

Research on the process of reflection-in-action: Here the emphasis is on stimulating and reinforcing practitioners to engage in reflective practice.

In recent years, the paradigm of applied science with its typical separation of responsibilities has been challenged also from another side. In his sociological studies of the ways technological tools (nuclear power stations, pesticides, vaccines, etc.) are developed, tested, and implemented and how these tools affect natural and social systems, the French philosopher Bruno Latour equally rejected the traditional separation between research and applications and introduced the concept of "collective experiment":

> In this new constellation, the expert is more and more disappearing. … The expert has been responsible for the mediation between the producers of knowledge and the society concerned with values and ends. However, in the collective experiments in which we are intrinsically caught up, exactly this separation of different roles has disappeared. So the position of the expert has been undermined. [It has] been proposed that the extinct concept of "expert" be replaced by the comprehensive concept of "co-researcher." (Latour 2001, p. 32, Translated by the author)

Obviously, the educational system is a "collective experiment in which we are intrinsically caught up." A separation between researchers who provide professional knowledge and teachers who simply use this knowledge is not appropriate.

Neither Schön's nor Latour's analyses provide practical solutions for mathematics education. However, they stimulate ideas for addressing the issue of managing complexity in this field.

2.4.3 Empowering Teachers to Cope with Systemic Complexity as Reflective Practitioners

In the first part of this section, proposals were made about how the collaboration between mathematics educators as reflective researchers and teachers as reflective practitioners can be filled with life. In the second, part these proposals are examined in the light of the preceding section.

A good general orientation for this section is given by John Dewey's view on the role teachers can play as "investigators." This view bears witness to the systemic sensibility of this farsighted author:

> It seems to me that the contributions that might come from classroom teachers are a comparatively neglected field; or, to change the metaphor, an almost unworked mine. ... There are undoubted obstacles in the way. It is often assumed, in effect if not in words, that classroom teachers have not themselves the training that will enable them to give effective intellectual cooperation. This objection proves too much, so much so that it is almost fatal to the idea of a workable scientific content in education. For these teachers are the ones in direct contact with pupils and hence the ones through whom the results of scientific findings finally reach students. They are the channels through which the consequences of educational theory come into the lives of those at school. I suspect that if these teachers are mainly channels of reception and transmission, the conclusions of science will be badly deflected and distorted before they get into the minds of pupils. I am inclined to believe that this state of affairs is a chief cause for the tendency, earlier alluded to, to convert scientific findings into recipes to be followed. The human desire to be an "authority" and to control the activities of others does not, alas, disappear when a man becomes a scientist. (Dewey 1929/1988, pp. 23–24)

As stated in Sect. 1, the core of mathematics education as a design science consists of the design, the empirical investigation, and the implementation of substantial learning environments with respect to boundary conditions set by society and beyond. So it has to be examined to determine in which way teachers can be enabled and encouraged to act as reflective practitioners in these three areas.

In terms of design: In the author's view the most important service mathematics educators can render to teachers is to provide them with elaborated substantial learning environments together with the structure-genetic didactical analyses on which the design has been based. The language in which substantial learning environments are communicated is understandable to teachers, so reflective practitioners have good starting points to transform what is offered to them into their context and to adapt, extend, cut, and improve it accordingly. In a recent paper, Chun Ip Fung has demonstrated teachers' creative work in this area by means of a striking example. He has shown that in this way a constructive dialogue between researchers and teachers can be established (see Sect. 3).

In terms of implementation: Individual learning environments and curricula cannot be implemented successfully without teachers' support. Implementation again requires teachers' creative powers in taking the local conditions into account and in adapting the proposed materials correspondingly. It is a triviality that teachers will engage more in the implementation of contents, objectives, or methods the more these are meaningful to them. Reflective researchers have to keep this in mind.

In terms of empirical evidence: This is a particularly important issue. In the author's view, teachers can best act as reflective investigators if empirical studies are attached to substantial learning environments and the results are communicable in a language that is understandable. Under these conditions teachers can cooperate in these studies and contribute to communicating the findings to practice.

However, empirical studies of the ordinary type are not the only way to get empirical evidence for the feasibility and the effectiveness of substantial learning environments. Another source is structure-genetic didactical analyses of the subject matter. Mathematics, well understood, provides not only the subject matter of teaching, but also methods of learning and teaching as it is itself the result of learning processes [see Sect. 4 in Wittmann 2015, with references to the fundamental paper by Dewey (1977)]. As these analyses imply empirical information on "staging" learning environments in the interaction with students, it is justified to call them empirical research "of the first kind," in distinction from ordinary empirical studies, the empirical research "of the second kind." Both structure-genetic didactical analyses and ordinary empirical studies are conducted either by researchers alone or determined by them. As teachers who collaborate with researchers in a research team are provided with additional information, have access to additional material, and enjoy support in various ways, they work under conditions that do not reflect the real practice. So for systemic reasons another type of empirical study seems promising: "collective teaching experiments." This empirical research "of the third kind" is obviously derived from Latour's "collective experiments." It is conducted by "freelancing" teachers in their daily practice, as will be discussed in some detail in the following section.

To conclude the present section, the above proposals for the interaction between reflective researchers and reflective teachers were examined against Schön's (1983) four types of "reflective research."

In terms of frame analysis: In order to provide teachers with an orientation beyond substantial learning environments, it is useful to summarize basic knowledge about mathematics, learning, and teaching mathematics in didactical principles. One principle, for example, is "orientation on fundamental mathematical ideas." This principle is based on Alfred N. Whitehead's view on mathematical education (Whitehead 1929), Jean Piaget's epistemology (e.g., Piaget 1972), and Hans Freudenthal's work, in particular Freudenthal (1983). This principle can be communicated to teachers best by linking it to series of learning environments in which this principle is a leading one.

In terms of repertoire-building research: Elaborated substantial learning environments form a repertoire for teaching par excellence. They contain the essential information for teaching. The reflective teacher, however, will not stick to this repertoire but use it as a springboard for exploring other learning environments.

In terms of research on fundamental methods of inquiry and overarching theories: In close connection with the two types of research discussed above, this type of research is directed toward introducing teachers into methods of inquiry inherent in mathematics and into elementary mathematical theories of subject matter that are relevant for teaching.

In terms of research on the process of reflection-in-action: The proposals that have been made for the design, empirical study, and implementation of a substantial learning environment are well suited to stimulating teachers to act as reflective practitioners.

It is obvious that both pre-service and in-service teacher education play a key role in educating reflective practitioners. Therefore the reflective mathematics educator is well advised to link his research to teacher education, including mathematics education and at least elementary mathematics.

2.4.4 Collective Teaching Experiments: A Joint Venture of Reflective Teachers and Reflective Researchers

The idea of encouraging teachers to become researchers of their own practice is not new at all. It has been particularly manifest in the Japanese tradition of lesson studies (Stigler and Hiebert 1999). In lesson studies, a group of teachers collaborates over a period of time on the design, empirical investigation, and implementation of learning environments. The lessons are given by teachers in actual classrooms, observed, discussed, and refined in several rounds until an acceptable result has been reached. A striking example is the recent Japanese research on elements of knot theory (Kawauchi and Yanagimoto 2012).

Collective teaching experiments are a modification of lesson studies in the following way: The reflective researchers offer research problems publicly and invite teachers to investigate them in their daily practice. There is only a loose connection with researchers, who collect the feedback and turn it into the improvement of the design and the implementation.

The following example illustrates this proposal:

In the past decades, German math teaching at the primary level has undergone a development away from standard procedures towards flexible strategies that reflect the true nature of mathematics. The curriculum developed in the project mathe 2000 is based on fundamental ideas of mathematics that can be developed over the grades. The arithmetical laws represent such a fundamental idea. The commutative and associative law of addition are implicitly introduced even at the kindergarten level and applied in a consequent and consistent manner at the primary and secondary level. The laws leave space for applying them in different ways, and it is

First plus tens, then plus ones		Tens plus tens, ones plus ones		Auxiliary problem	
$\underline{47 + 35 = 82}$	$\underline{65 + 28 = 93}$	$\underline{47 + 35 = 82}$	$\underline{65 + 28 = 93}$	$\underline{47 + 35 = 82}$	$65 + 28 = 93$
$47 + 30 = 77$	$65 + 20 = 85$	$40 + 30 = 70$	$60 + 20 = 80$	$45 + 35 = 80$	$63 + 30 = 93$
$77 + 5 = 82$	$85 + 8 = 93$	$7 + 5 = 12$	$5 + 8 = 13$		

Fig. 13 Basic strategies for adding two-digit numbers

First minus tens, then minus ones		Tens minus tens, ones minus ones		Auxiliary problem	
$\underline{87 - 35 = 52}$	$\underline{65 - 28 = 37}$	$\underline{87 - 35 = 52}$	$\underline{65 - 28 = 37}$	$\underline{87 - 35 = 52}$	$65 - 28 = 37$
$87 - 30 = 57$	$65 - 20 = 45$	$80 - 30 = 50$	$60 - 20 = 40$	$85 - 35 = 50$	$\underline{65 - 25 = 40}$
$57 - 5 = 52$	$45 - 8 = 37$	$7 - 5 = 2$	$5 - 8 = -3$		$40 - 3 = 37$

Fig. 14 Basic strategies for subtracting two-digit numbers

important that teachers and children become aware of this freedom by being offered different strategies.

In adding two digit numbers there are essentially three basic strategies. None of them causes problems (see Fig. 13).

All three strategies can be transferred to subtraction. However, the second strategy causes a problem when the ones in the subtrahend exceed the ones in the minuend (see Fig. 14).

Experience shows that many children transfer the strategy "tens plus tens, ones plus ones" blindly to the strategy "tens minus tens, ones minus ones" and arrive at wrong results. For the problem $65 - 28$, for example, they calculate $60 - 20 = 40$, $8 - 5 = 3$ and get 43. So many teachers, supported by many textbook authors, reject and avoid this strategy either by prescribing the first subtraction strategy or by modifying the critical one as follows:

$50 - 20 = 30$, $15 - 8 = 7$, so $65 - 28 = 37$. For us such didactic compromises are no option. We believe that it is better not to avoid the critical strategy also for its long-term importance. As early as 1977, the Dutch computer scientist Sytze van der Meulen, after his talk in our colloquium at the IEEM (Institute for Development and Research in Mathematics Education) in Dortmund, left a message in our guestbook that has since been a continuous reminder to us:

When a boy answers the question "how much is $7 - 4$" with 3, he is not a genius when his age is 7. When this boy answers the question how much is "$4 - 7$" with "there are three missing" he shows some intelligence, but still is not a genius at the age of 7. The tragedy of our school-education is that this boy at the age of 11 may have difficulties with the concept of negative numbers. The tragedy of his teacher is that he missed 4 years of the boy's development!

Over the years we have taken several steps to overcome teachers' scruples concerning the subtraction strategy "tens minus tens, ones minus ones," and we have stimulated teaching experiments on a small scale. Since 1995 we have been using any opportunity to explain this strategy to teachers and to ask them to try it out with their students.

We recommend explaining $5 - 8 = -3$ as follows:

> We have 5 and have to take away 8. First we take away 5, and then we have to take away 3 more. In order not to forget this, we note it down as "−3." Finally we take away 3 by breaking up one ten into 10 ones, and remove 3 of them.

By means of bars of ten and counters, this procedure can be well demonstrated step by step.

We also tell teachers that this strategy has an important advantage: The calculations are easier in comparison with the first strategy, so this strategy seems particularly suited for weak students despite the first impression that it might not be appropriate for them.

One teacher did a small study and communicated it to us: After she had taught subtraction in the hundreds space in the traditional way without the critical strategy, she administered a test to her class. Then she introduced this strategy and repeated the test. It turned out that the results were no better and no worse. As she had a class with many weak students who had difficulties with this strategy, her recommendation was to avoid this strategy. Nevertheless, we continued our "propaganda" for this strategy and improved our proposal on how to explain it to students.

One year later an "unforced" e-mail arrived from this teacher that read as follows:

> Last year my students did have difficulties with the strategy "tens minus tens, ones minus ones" because of the negative numbers. Now I would like to report on my latest experiences with this strategy in Grade 3. In the introductory lesson, I wrote down the problem $629 - 263$ without any repetition of the calculations from the year before and without any explanations from my side. Apart from very few exceptions the children calculated $20 - 60 = -40$. For them it was obvious: "The result is −40, exactly as it was last year with the ones." I would emphasize that my class is not a superclass, and that I have very many weak students. For them, calculations with negative numbers do not cause any problem. I am strongly in favour of introducing this strategy already in Grade 2.

As many teachers have still reservations about this strategy, we have refined our explanation. We recommend now to let students distinguish between the cases when there are enough ones in the minuend and those where more ones have to be taken away than are available. We recommend proposing packages of subtraction problems to students and asking them to mark those where a minus sign appears in the ones calculation with an asterisk before they perform the actual calculations. The latest improvement in teaching this strategy is to explicate $5 - 8 = -3$ in more detail: $5 - 5 - 3 = 0 - 3 = -3$. We do not have enough feedback from teachers as this moment; however, we are confident that this step will increase the acceptance of this strategy.

These experiences and similar ones with other issues have led us to a far-reaching conclusion: Our main publication, a handbook for teaching arithmetic at the primary level (Wittmann and Müller 1990, 1992) will be rewritten soon with the explicit invitation to teachers to conduct collective teaching experiments. All learning environments collected in this new book will belong to the standard curriculum. They will be accompanied not only with the general recommendation to read them critically and to test them in their classroom but also to participate in conducting collective teaching experiments in cooperation with other teachers. We will also create a platform for an exchange about the experiences with these experiments.

Issues that are of particular interest for us are operative proofs, the use of our course on mental arithmetic, and the use of new digital means of representation.

2.4.5 Closing Remarks: The Role of Mathematics in Mathematics Education

It is important to realize that the research and development program that has been described in this chapter heavily depends on resources that are offered by "well-understood" mathematics. "Well-understood" means that mathematics is seen as a social organism that has developed in history and it still developing with strong relations to many areas of human life, and that also the mathematical knowledge of the individual is seen as an organism in its genesis from tiny seeds to a more or less extensive body. Doing mathematics is learning mathematics and learning mathematics should also be firmly linked to doing mathematics. Therefore, the interaction between teachers and students and between teachers and researchers can greatly profit from relying on the adaptability of elementary mathematical structures with respect to students' individual cognitive levels and on the processes inherent in vital mathematics.

When once asked what his motives as a mathematician were for engaging in mathematics education Hans Freudenthal replied: "I want to understand better what mathematics is about." The reverse also holds: mathematics educators who want to understand better what mathematics education should be about are well advised to study elementary mathematical structures thoroughly. It is highly rewarding to "unfreeze" the educational material that is "deep-frozen" in polished presentations of mathematics, as they are common in higher mathematics. After all "well-understood" mathematics is the best common reference for all involved in teaching and learning mathematics: researchers, teachers, and students. "Theories of mathematics education" like those collected in Sriraman and English (2010) are far from being suited for establishing a systemic cooperation between reflective researchers and reflective teachers.

2.5 Design Science and Design Research: Commonalities and Variations

Bettina Rösken-Winter and Marcus Nührenbörger

Having explored the characteristics of design sciences in detail, and also against the background of those days when Wittmann contributed his thoughts, we now point out current trends both in Germany and from an international perspective.

2.5.1 Developments in the Field of Design Science in Germany

In Germany, the concept of design science has been further developed at the Institute for Development and Research of Mathematics Teaching founded by Wittmann. As in the mathe 2000 project, the research group at the institute has focused on constructive didactical developmental work and reconstructive empirical research of the specific features, conditions, and structures of (case-related) teaching and learning processes. The interplay of constructive developmental work, empirical research, and theory development has been reflected in the various projects of the institute. Studying the learning conditions and processes of the specific mathematical content dependent on the various school levels involves exploring students' process-related skills, the individual characteristics of the learning process, and the dynamic processes taking place in the classroom. Hence, the alignment of the research group is still strongly oriented towards Wittmann's conceptualisation of design science. However, the aforementioned particularities also document a shift from design science to design research, as Prediger (2013) pointed out: "In our research group, we follow the programme of Didactical Design Research as formulated by Gravemeijer and Cobb (2006), which combines the concrete design of learning arrangements with fundamental research on the initiated learning processes" (p. 4).

The Dortmund design research approach pursued in the project FUNKEN is representative of the current developments in design science in Germany. On the one hand, the FUNKEN goal has been to produce teaching and learning arrangements in different subject areas on the basis of epistemological analyses. On the other hand, the learning processes initiated by the teaching and learning arrangements have been studied in a cycle of iterative, closely linked steps. In this sense, Wittmann's structure-genetic didactical analyses are related to international design research concepts (cf. Van den Akker et al. 2006; Cobb et al. 2003) and specifically to those favouring process-oriented analyses of learning (Gravemeijer and Cobb 2006).

Gravemeijer and Cobb (2006) characterize their learning perspective by three phases of conducting a design experiment: "(1) preparing for the experiment, (2) experimenting in the classroom, and (3) conducting retrospective analyses" (p. 53). The Dortmund design research approach comprises of four phases that are closely related to each other and can run iteratively several times (Fig. 15).

The model consists of four distinct steps distributed equally over two levels, one presenting the design level and one the research level. Thus, the two levels can be differentiated as being constructive or reconstructive by nature. Ultimately, they reflect the two archetypes of design research: the former targets towards direct practical use and the latter towards theory development with respect to both the teaching and learning processes involved. The starting point of the research process generally lies in the field of developing new teaching and learning arrangements (top half of the model).

Within the model, specifying and structuring the learning objects serves as the starting point of object-oriented didactical developmental work. Therefore, before

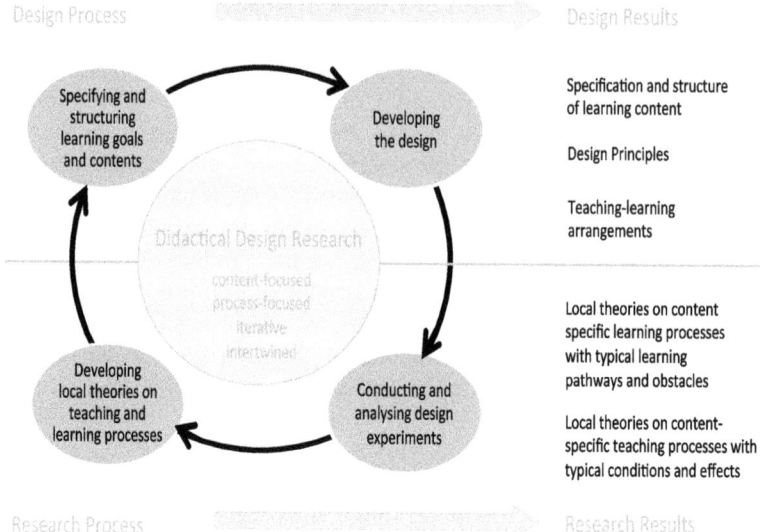

Fig. 15 Four phases of didactical design research (Prediger et al. 2012, translated in Prediger and Zwetzschler 2013)

initially creating a teaching and learning arrangement, a theory-based analysis of the learning object is performed using epistemological and didactic perspectives. Structuring the learning object includes the identification of the relevant contexts, including basic and appropriate means of representation, and finally sequencing the content appropriately, taking into account the student viewpoint. Here, the elementary didactical design principles, as formulated by Wittmann, are always decisive. The next step is dedicated to developing a design to be tested in practice. Therefore, it is essential to include a didactical theory on which design decisions can be based; e.g., the importance of presentation changes, forms of productive practice, support in case of learning problems, or initiating social learning.

Implementing and evaluating design experiments aims at studying learning processes and their initiation by specific design features. For this purpose (usually multiple) sample trials are performed with learners in various constellations (e.g., in an individual clinical interview, in group interviews, or in the classroom under real school conditions). Particularly in laboratory-like situations, the researcher works as a teacher and is responsible for motivating and supporting the students to trigger specific thinking processes that ultimately provide information on students' approaches (see also Cobb et al. 2003).

Lastly, the fourth step consists of "elaborating on an empirically grounded *subject-specific local instruction theory* that specifies the epistemological structure of the particular learning content" (Prediger 2013, p. 4). This includes possible courses, typical obstacles in the learning process, conditions and effects of exercises, and the means of support used. The results of the reconstructive analyses form the new and enhanced basis for the next iteration in the developmental research cycle and allude to required changes of the learning object sequencing. Ultimately, empirical findings only become theory if it is possible to extract the analysed phenomena and patterns of their very specific context of origin and to discern the generalisable parts through comparison of cases. Nevertheless, the gained learning theory remains local in nature, because it is always closely linked to the original context of the case study and the specific learning object.

Currently, the design research approach favoured in the Dortmund research group serves as a role model to conduct design research in the field of mathematics teachers' professional development. Within the German Centre for Mathematics Teacher Education (DZLM), design research is conceptualised to investigate course designs and to spread innovations on a large scale (cf. DZLM 2015; Roesken-Winter and Szczesny in press).

2.5.2 Developments of Design Science from an International Perspective

While Wittmann had developed his view of mathematics education as a design science since the 1970s in Germany and implemented the developmental research project mathe 2000, other trends in mathematics education were appearing both based on this and in parallel. In a similar manner, they placed the design of teaching examples at the core of mathematics research. However, the approaches differed in their conceptualizations and bore different names, such as design experiments (e.g., Brown 1992), design-based research (e.g., Barab and Squire 2004), developmental research (e.g., Gravemeijer 1994), or engineering research (e.g., Burkhardt 2006; Lesh and Sriraman 2005). All these concepts agree in that they are theoretically sound while remaining pragmatic. That is, as regards practice, teaching examples are developed and then tested in the classroom, and design principles are derived. In addition, theoretical aspects are involved as the effects of implementing specific teaching examples are investigated and the conditions for substantial learning and related teaching processes analysed. Recently, different calls have emerged to synthesise the many different facets of design research. Van den Akker et al. (2006), for instance, demanded guiding principles for scientific research:

- pose significant questions that can be investigated
- link research to relevant theory
- use methods that permit direct investigation of the question
- provide a coherent and explicit chain of reasoning
- replicate and generalize across studies
- disclose research to encourage professional scrutiny and critique (p. 17).

Accordingly, Prediger et al. (2015) advocated the following five key features as characteristics for design research:

- Interventionist: The intent is to create and study new forms of instruction; in this sense, it must be interventionist rather than naturalistic.
- Theory generative: The goal is to generate theories (to develop new theory concepts or to refine theories) about the process of learning and the means of supporting learning processes.
- Prospective and reflective: The connection between theory and experiment is twofold: prospective and reflective. Theory prospectively informs the design for the design experiment and is further developed in the retrospective reflection on deviances between the expected and the observed teaching and learning processes
- Iterative: Design research studies concern different iterative cycles (micro-design and macro-design cycles) of invention and revision. The micro-design cycles take place within one design experiment when the researchers try to adapt both the instructional activities and the theory that underpins them. Macro-design cycles occur when design experiments are repeated while building on one or more preceding design experiments.
- Ecologically valid and practice-oriented: When design research is situated within real classrooms, the conditions of the study already represent the complexity of conditions of practice.

Conducting design research to enable innovations in mathematics education has been extended to various fields of research. The focus currently is on not only modelling student and teacher learning but also on broader approaches that aim at spreading programmes or projects at large (cf. Kelly et al. 2008). Compared to the initial aim of establishing mathematics education as a design science, conceptualizing design research draws on different issues. More emphasis is placed on structural features that distinguish the approach from common methodologies aligned to a qualitative or quantitative research paradigm. However, both approaches strongly advocate balancing theory and practice and measuring empirical research in terms of its practical relevance.

3   Summary and Looking Ahead

In this survey we explored how and why one prominent line of research elaborates on conceptualizing mathematics education as a design science:

- Particularly, we reflected on Wittmann's contribution to establishing such a viewpoint of mathematics education research as a field of scientific inquiry.
- Developments within the German context were reflected against the background of European strands of similar orientations such as in France and the Netherlands.
- Specific attention was paid to the role of substantial learning environments with respect to both designing and researching them.

- Three examples were presented that depart from design science but put emphasis on different accentuations as they reflect current trends in mathematics education research. One example contributed additional reflections on the role of research related to substantial learning environments and how research and practice can be brought together, still questioning the role of empirical research in mathematics education.
- Finally, design science was reflected in view of national and international trends, revealing essential characteristics that place, for instance, the design research approach as continuous development.

As Wittmann pointed out in 1998, mathematics education as design science can take different paths. Particularly, he underlined that different facets of design science that emerge at the same time can be valued as a sign of progress, as far as they pursue the main goal to relate design and empirical research. "In order to develop mathematical didactics as design science, it is important to find ways of connecting design and empirical research" (Wittmann 1998, p. 337; translated by the authors).

Current discussions underline a shift from the need to establish mathematics education as a scientific field in its own right to differentiating apparently distinct design research approaches. However, design research has permeated the field substantially and many research contributions focus on investigating the design and re-design of learning environments either with respect to children's or teachers' learning. In 2006, Burkhardt reminded us of the following:

> Educational research still has much less impact on policy and practice than we would wish. If politicians have a problem in their education system, is their first move to call a research expert? Not often. Indeed, in most countries, there is no obvious link between changes in practice and any of the research of the tens of thousands of university researchers in education around the world. (Burkhardt 2006, p. 122)

It is the very nature design science and design research to balance theory and practice fruitfully. Researchers and teachers work together to promote mathematics learning and to opt for changes that ensure substantial developments. Departing from such an understanding, the main challenges lie in keeping the core of the research approach when adapting methods to the specific requirements at a particular time and scaling innovations at large.

References

Adler, J., & Davis, Z. (2006). Opening another black box: Researching mathematics for teaching in mathematics teacher education. *Journal for Research in Mathematics Education, 37*(4), 270–296.

Artigue, M. (1994). Didactical engineering as a framework for the conception of teaching products. In R. Biehler, R. W. Scholz, R. Sträser, & B. Winkelmann (Eds.), *Didactics of mathematics as a scientific discipline* (pp. 27–39). Dordrecht: Kluwer.

Ball, D. L., & Bass, H. (2000). Interweaving content and pedagogy in teaching and learning to teach: Knowing and using mathematics. In J. Boaler (Ed.), *Multiple perspectives on mathematics teaching and learning* (pp. 83–104). Westport, CT: Ablex Publishing.

Barab, S., & Squire, K. (2004). Design-based research: Putting a stake in the ground. *Journal of the Learning Sciences, 13*(1), 1–14.

Brousseau, G. (1997). *Theory of didactical situations in mathematics*. Dordrecht: Kluwer.

Brown, A. L. (1992). Design experiments: Theoretical and methodological challenges in creating complex interventions in classroom settings. *Journal of the Learning Sciences, 2*(2), 141–178.

Burkhardt, H. (2006). From design research to large-scale impact: Engineering research in education. In J. van den Akker, K. Gravemeijer, S. McKenney, & N. Nieveen (Eds.), *Educational design research* (pp. 121–150). London: Routledge.

Chevallard, Y. (1991). *La transposition didactique: Du savoir savant au savoir enseigné*. La pensé sauvage: Grenoble.

Chevellard, Y., & Sensevy, G. (2014). Anthropological approaches in mathematics education, French perspectives. In S. Lerman (Ed.), *Encyclopedia of Mathematics Education* (pp. 38–43). Dordrecht, Heidelberg, New York, London: Springer.

Chow, W.-Y. (2006). Do pupils really know "factor" (in Chinese)? *EduMath, 22*, 25–30.

Cobb, P., Confrey, J., diSessa, A., Lehrer, R., & Schauble, L. (2003). Design experiments in educational research. *Educational Researcher, 32*(1), 9–13.

Conference Board of the Mathematical Sciences, A. M. S. (2001). *The mathematical education of teachers*. Providence, R.I.: American Mathematical Society.

Davis, B., & Simmt, E. (2006). Mathematics-for-Teaching: An ongoing investigation of the mathematics that teachers (need to) know. *Educational Studies in Mathematics, 61*(3), 293–319.

Design-Based Research Collective. (2003). Design-based research: An emerging paradigm for educational inquiry. *Educational Researcher, 32*(1), 5–8.

Dewey, J. (1904/1977). The relation of theory to practice in Education. In J. A. Boydston (Eds.), *Dewey, J., The middle works 1899–1924*, (Vol. 3, pp. 249–272). Carbondale, Illinois: SIZ Press.

Dewey, J. (1929/1988). The sources of a science of education. In J. A. Boydston (Eds.), *Dewey, J., The later works 1925–1953* (Vol. 5, pp. 1–40). Carbondale, Illinois: SIU Press.

DZLM. (2015). *Theoretischer Rahmen des Deutschen Zentrums für Lehrerbildung Mathematik*. http://www.dzlm.de/files/uploads/DZLM_Theorierahmen.pdf. Accessed 30 March 2015.

Fletcher, P. (1964). *Some lessons in mathematics*. London: CUP.

Freudenthal, H. (1973). *Mathematics as an educational task*. Dordrecht: Reidel.

Freudenthal, H. (1983). *Didactical phenomenology of mathematical structures*. Dordrecht: Reidel.

Freudenthal, H. (1991). *Revisiting mathematics education: China lectures*. Dordrecht: Kluwer Academic Publishers.

Fung, C.-I. (2004). How history fuels teaching for mathematising: Some personal reflections. *Mediterranean Journal for Research in Mathematics Education, 3*(1–2), 125–146.

Fung, C.-I. (2008). The teaching of division with remainder in elementary school (in Chinese). *EduMath, 27*, 34–46.

Gravemeijer, K. (1994). *Developing realistic mathematics education*. Utrecht: Cd-ß Press.

Gravemeijer, K. (1999). How emergent models may foster the constitution of formal mathematics. *Mathematical Thinking and Learning, 1*(2), 155–177.

Gravemeijer, K., & Cobb, P. (2006). Design research from a learning design perspective. In J. van den Akker, K. Gravemeijer, S. McKenney, & N. Nieveen (Eds.), *Educational design research* (pp. 17–51). London: Routledge.

Hanna, G. (2000). Proof, explanation and exploration: An Overview. *Educational Studies in Mathematics, 44*, 5–23.

Heath, T. L. (1956). *The thirteen books of Euclid's Elements* (2nd ed.). New York: Dover Publications.

Hefendehl-Hebeker, L. (1998). The practice of teaching mathematics: Experimental conditions of change. In F. Seeger, J. Voigt, & U. Waschescio (Eds.), *The culture of mathematics classrooms* (pp. 104–126). Cambridge: University Press.

Heid, M. K., Middleton, J. A., Larson, M., Gutstein, E., Fey, J. T., King, K., & Tunis, H. (2006). The challenge of linking research and practice. *Journal for Research in Mathematics Education, 37*(2), 76–86.

Heintz, B. (2000). *Die Innenwelt der Mathematik. Zur Kultur und Praxis einer beweisenden Disziplin.* Springer Verlag: Wien.

Heuvel-Panhuizen, M. (2003). The learning paradox and the learning miracle: Thoughts on primary school mathematics education. *Journal für Mathematik-Didaktik, 24*(2), 28–53.

Kawauchi, A., & Yanagimoto, T. (Eds.). (2012). *Teaching and learning knot theory in school mathematics.* Tokyo, Heidelberg: Springer.

Kelly, A., Lesh, R., & Baek, J. (Eds.). (2008). *Handbook of design research methods in education.* London: Routledge.

Kennedy, K. M. M. (1997). The connection between research and practice. *Educational Researcher, 26*(7), 4–12.

Kilpatrick, J. (2008). The development of mathematics education as an academic field. In M. Menghini, F. Furinghette, L. Giacardi, & F. Azarello (Eds.), *The first century of the International Commission on Mathematical INstruction (1908–2008): Reflecting and shaping the world of mathematics education* (pp. 25–39). Rome, Italy: Istituto della encyclopedia Italiana.

Krummheuer, G. (1995). The ethnography of argumentation. In P. Cobb & H. Bauersfeld (Eds.), *The emergence of mathematical meaning: interaction in classroom cultures* (pp. 229–270). Hillshale, N.J.: Lawrence Erlbaum.

Latour, B. (2001). Ein Experiment von und mit uns allen. *DIE ZEIT* (16), 31–32. (Resource document. http://www.zeit.de/2001/16/Ein_Experiment_von_und_mit_uns_allen).

Latour, B. (2011). From multiculturalism to multinaturalism: What rules of method for the new socio-scientific experiments? *Nature and Culture, 6*(1), 1–17.

Lesh, R., & Sriraman, B. (2005). Mathematics education as a design science. *ZDM, 37*(6), 490–505.

Link, M. (2012). *Grundschulkinder beschreiben operative Zahlenmuster.* Wiesbaden: Springer.

Leitzel, J. R. C. (1991). *A call for change: Recommendations for the mathematical preparation of teachers of mathematics.* Washington, DC: Mathematical Association of America.

Margolinas, C., & Drijvers, P. (2015). Didactical engineering in France; an insider's and an outsider's view on its foundations, its practice and its impact. *ZDM Mathematics Education, 47*(6), 893–903.

Malik, F. (1986). *Strategie des Managements komplexer Systeme.* Bern: Haupt.

Miller, M. (1986). *Kollektive Lernprozesse.* Frankfurt: Suhrkamp.

Miller, M. (2002). Some theoretical aspects of systemic learning. *Sozialer Sinn, 3*, 379–422.

Nührenbörger, M., & Schwarzkopf, R. (2010). Diskurse über mathematische Zusammenhänge. In C. Böttinger, K. Bräuning, M. Nührenbörger, R. Schwarzkopf, & E. Söbbeke (Eds.), *Mathematik im Denken der Kinder* (pp. 169–215). Seelze: Kallmeyer.

Nührenbörger, M., & Schwarzkopf, R. (2016). Processes of mathematical reasoning of equations in primary mathematics lessons. In N. Vondrová (Ed.), *Proceedings of the 9th Congress of the European Society for Research in Mathematics Education (CERME 9)* (pp. 316–323). Prag: ERME.

Nührenbörger, M., & Steinbring, H. (2009). Forms of mathematical interaction in different social settings—examples from students', teachers' and teacher-students' communication about mathematics. *Journal of Mathematics Teacher Education, 12*(2), 111–132.

Piaget, J. (1972). *Theorien und Methoden der modernen Erziehung.* Wien, München: Molden.

Piaget, J. (1985). *The equilibration of cognitive structure: the central problem of intellectual development.* Chicago, IL: University of Chicago Press.

Prediger, S. (2013). Focussing structural relations in the bar board—a design research study for fostering all students' conceptual understanding of fractions. In B. Ubuz, C. Haser, & M. A. Mariotti (Eds.), *Proceedings of the 8th Congress of the European Society for Research in Mathematics Education* (pp. 343–352). Ankara: METU University.

Prediger, S., & Zwetzschler, L. (2013). Topic-specific design research with a focus on learning processes: The case of understanding algebraic equivalence in grade 8. In T. Plomp & N. Nieveen (Eds.), *Educational design research: Illustrative cases* (pp. 407–424). Enschede: SLO, Netherlands Institute for Curriculum Development.

Prediger, S., Link, M., Hinz, R., Hußmann, S., Thiele, J., & Ralle, B. (2012). Lehr-Lernprozesse initiieren und erforschen - Fachdidaktische Entwicklungsforschung im Dortmunder Modell. *Mathematischer und Naturwissenschaftlicher Unterricht, 65*(8), 452–457.

Prediger, S., Gravemeijer, K., & Confrey, J. (2015). Design research with a focus on learning processes—an overview on achievements and challenges. *ZDM, 47*(6), 877–891.

Roesken-Winter, B., & Szczesny, M. (in press). Continuous professional development (CPD): Paying attention to requirements and conditions of innovations. In S. Doff, & R. Komoss (Eds.), *How does change happen? Wandel im Fachunterricht analysieren und gestalten.* Berlin, Heidelberg, New York: Springer.

Ruthven, K., Laborde, C., Leach, J., & Tiberghien, A. (2009). Design tools in didactical research: Instrumenting the epistemological and cognitive aspects of the design of teaching sequences. *Educational Researcher, 38*(5), 329–342.

Schön, D. A. (1983). *The reflective practitioner. How professionals think in action.* New York: Basic books.

Schön, D. A. (1991). *The reflective turn: Case studies in and on educational practice.* New York: Teachers College (Columbia).

Schwarzkopf, R. (2000). Argumentation processes in mathematics classrooms—Social regularities in argumentations processes. In GDM (Eds.), *Developments in Mathematics Education in Germany—Selected Papers from the Annual Conference on Didactics of Mathematics Potsdam*, 139–151.

Schwarzkopf, R. (2003). Begründungen und neues Wissen: Die Spanne zwischen empirischen und strukturellen Argumenten in mathematischen Lernprozessen der Grundschule. *Journal für Mathematik-Didaktik, 24*(3/4), 211–235.

Skovsmose, O. (2001). Landscapes of investigation. *ZDM: Zentralblatt für Didaktik der Mathematik, 33*(4), 123–132.

Skovsmose, O. (2011). *An invitation to critical mathematics education.* Rotterdam: Sense.

Sfard, A. (2005). What could be more practical than good research? *Educational Studies in Mathematics, 58*(3), 393–413.

Simon, H. A. (1970). *The sciences of the artificial.* Cambridge, MA: MIT Press.

Sriraman, B., & English, L. (Eds.). (2010). *Theories of mathematics education. Seeking new frontiers.* Berlin Heidelberg: Springer.

St. Clair, R. (2005). Similarity and superunknowns: An essay on the challenges of educational research. *Harvard Educational Review, 75*(4), 435–453.

Steinbring, H. (2005). *The construction of new mathematical knowledge in classroom interaction.* Berlin: Springer.

Steinweg, A. S. (2006). Sich ein Bild machen. *Terme und figurierte Zahlen. mathematik lehren, 136*, 14–17.

Steinweg, A. S. (2013). *Algebra in der Grundschule.* Berlin: Springer.

Stigler, J., & Hiebert, J. (1999). *The teaching gap.* New York: The Free Press.

Voigt, J. (1994). Negotiation of mathematical meaning and learning mathematics. *Educational Studies in Mathematics, 26*(2–3), 275–298.

Van den Akker, J., Gravemeijer, K., McKenney, S., & Nieveen, N. (2006). Introducing educational design research. In J. van den Akker, K. Gravemeijer, S. McKenney, & N. Nieveen (Eds.), *Educational design research* (pp. 3–7). London: Routledge.

Van den Heuvel-Panhuizen, M., & Drijvers, P. (2014). Realistic mathematics education. In S. Lerman (Ed.), *Encyclopedia of mathematics education* (pp. 521–525). Dordrecht: Springer.

Watson, A. (2008). School mathematics as a special kind of mathematics. *For the Learning of Mathematics, 28*(3), 3–7.

Wheeler, D. (1967). *Notes on primary mathematics.* London: CUP.

Whitehead, A. N. (1929). *The aims of education and other essays.* New York: Macmillan.

Wieringa, N. (2011). Teachers' educational design as process of reflection-in-action. *Curriculum Inquiry, 42*(1), 167–174.

Winter, H. (1982). Das Gleichheitszeichen im Mathematikunterricht der Primarstufe. *Mathematica Didacta, 5,* 185–211.

Wittmann, E Ch. (1992). Mathematikdidaktik als "design science". *Journal für Mathematikdidaktik, 13,* 55–70.

Wittmann, E. Ch. (1995). Mathematics education as a "design science". *Educational Studies in Mathematics 29,* 355–374 [repr. In A. Sierpinská, & J. Kilpatrick (Eds.) (1998), *Mathematics Education as a Research Domain. A Search for Identity* (pp. 87–103). Dordrecht: Kluwer].

Wittmann, E. Ch. (1998). Operative proof in elementary school. *Mathematics in School, 27*(5).

Wittmann, E. Ch. (2001a). Developing mathematics education in a systemic process. *Educational Studies in Mathematics, 48*(1), 1–20.

Wittmann, E. Ch. (2001b). Drawing on the richness of elementary mathematics in designing substantial learning environments. In M. van den Heuvel-Panhuizen (Ed.), *Proceedings of the 25th Conference of pme* (pp. 193–197), Vol. 1, Utrecht Netherlands.

Wittmann, E. Ch. (2004). *Empirical research centred around substantial learning environments.* (unpublished Plenary Lecture delivered at the Annual Meeting of the Japanese Society of Mathematics Education, Okayama, 20–22 November 2004).

Wittmann, E. Ch. (2015). Structure-genetic didactical analyses—Empirical research "of the first kind". In P. Błaszczyk, B. Pieronkiewicz, & M. Samborska (Eds.), *Mathematical Transgressions 2015* (pp. 5–19). Kraków: PWN.

Wittmann, E. Ch., & Müller, G. N. (1990). *Handbuch produktiver Rechenübungen* (Vol. 1). Stuttgart: Klett.

Wittmann, E. Ch., & Müller, G. N. (1992). *Handbuch produktiver Rechenübungen* (Vol. 2). Stuttgart: Klett.

Wittmann, E. Ch., & Müller, G. N. (1988). When is a proof a proof? *Bulletin of Social Mathematics in Belgium, 1,* 15–40.

Wittmann, E. Ch., & Müller, G. N. (2012). *Das Zahlenbuch.* Leipzig: Klett.

Yackel, E., & Cobb, P. (1996). Sociomathematical norms, argumentation, and autonomy in mathematics. *Journal for Research in Mathematics Education, 27*(4), 458–477.

Further Reading

Prediger, S., Gravemeijer, K., & Confrey, J. (2015). Design research with a focus on learning processes *ZDM Mathematics Education, 47*(6).

Van den Akker, J., Gravemeijer, K., McKenney, S., & Nieveen, N. (Eds.). (2006). *Educational design research.* London: Routledge.

Wittmann, E. Ch. (1995). Mathematics education as a "design science". *Educational Studies in Mathematics, 29,* 355–374 [repr. in A. Sierpinská, & J. Kilpatrick (Eds.) (1998), *Mathematics Education as a Research Domain. A Search for Identity* (pp. 87–103). Dordrecht: Kluwer].

Wittmann, E. Ch. (2001). Developing mathematics education in a systemic process. *Educational Studies in Mathematics, 48*(1), 1–20.

www.ingramcontent.com/pod-product-compliance
Ingram Content Group UK Ltd.
Pitfield, Milton Keynes, MK11 3LW, UK
UKHW020216231225
466357UK00011B/174